Project

Management

How to Become a Better Project
Manager

*(How to Effectively Manage Any Project Like a
True Professional)*

Barry Westberry

Published By **Regina Loviusher**

Barry Westberry

Project Management: How to Become a Better Project Manager (How to Effectively Manage Any Project Like a True Professional)

ISBN   978-1-998901-96-8

No part of this guidebook shall be reproduced in any form without permission in writing from the publisher except in the case of brief quotations embodied in critical articles or reviews.

Legal & Disclaimer

The information contained in this book is not designed to replace or take the place of any form of medicine or professional medical advice. The information in this book has been provided for educational & entertainment purposes only.

The information contained in this book has been compiled from sources deemed reliable, and it is accurate to the best of the Author's knowledge; however, the Author cannot guarantee its accuracy and validity and cannot be held liable for any errors or omissions. Changes are periodically made to this book. You must consult your doctor or get professional medical advice before using any of the suggested remedies, techniques, or information in this book.

Table Of Contents

Chapter 1: It's All About You

A Project Manager's Self-Awareness

Putting your self first isn't constantly clean but, it's miles essential whilst studying a brand new expertise or belief method. Understanding the manner you react to others and the situations you are in is the first step to growing your project manipulate emotional intelligence.

Several years ago I become going for walks alongside undertaking supervisor, Matt, who was numerous years my senior at the equal time as it came to mission manage enjoy. I seemed as plenty as him and quite frankly, desired to be similar to him. That is, till, this one assignment.

The task we labored on modified right right into a multi-year infrastructure project and required the opportunity of loads of networking components. We had groups from everywhere in the worldwide and in almost every us of a in the United States. The try end up very huge and complex.

As we began out the undertaking Matt changed into on top of factors. You ought to tell he have turn out to be very skilled and knew a manner to run a project. His documentation grow to be prepared, he understood what to do in the starting tiers and he successfully communicated the expectancies of the challenge to our crew at some point of the venture kickoff.

Several weeks into our efforts, the economic corporation decided on a change in direction. Due to a charge variety scarcity, the project timeline had to be shortened so as for the project to have specific sufficient funding. While this isn't always uncommon, it does add pressure to on the venture supervisor but in addition to the complete Even the maximum pro of challenge managers can cope with the changing of a previously agreed upon price range and scope, however only for good-bye.

About months after the preliminary discovery of price variety constraints, you can inform the strain modified into beginning to mount on Matt. His frustration modified into palpable to the relaxation of our crew. This changed into in particular real in our conferences with each the center crew in addition to people with vendors and business stakeholders. He turned into without problem agitated, fantastically irritated and, like an open wound, fantastically willing. Everyone found that he changed into starting to get flustered an awful lot faster and much less complicated now. Where he was calm as mild tides rolling in, he modified into now emotionally volatile, crashing like typhoon waves. Symptoms of the adjustments in the assignment precipitated delays inside the device delivery and due to finances cuts, team people were given introduced paintings from exceptional duties. This supposed they did no longer have the same quantity of time to decide to our task as they did at the same time as we began out out.

In retrospect, I now apprehend that I became witnessing Matt's specific degree of emotional intelligence in motion. His emotional intelligence "playbook" if you can. The greater pressure Matt changed into underneath the much less complex it have emerge as to take a look at the pressure in his face, listen it in his voice, and characteristic a examine it in his interactions with others. Matt's low diploma of emotional intelligence did now not equip him to be self-aware about his actions or to recognize his deteriorating notion many of the organization and others.

So, as a project manager, what are you capable of do to grow to be extra self-conscious and consequently boom your emotional intelligence? Let's take a look at the subsequent three strategies you can growth yourself-attention.

1) Assess Yourself

You are your very own greatest fine buddy and asset so discover time for a self-assessment. Take time to make your self acquainted collectively along with your strengths and weaknesses. This is not supposed as an workout as a manner to select yourself harshly, or pat yourself at the lower back. This workout is surely as a manner to compare in which you are. The answers you display are neither pinnacle nor lousy, they truely are so that you can hold that mirror as plenty as yourself and in fact be honest.

Next, in yourself-assessment, you may want to understand your triggers. What are the ones subjects that reason you to get upset, lose manipulate, or get sidetracked? Are you an emotional elevator? Admittedly, grasping an notable knowledge of my very personal triggers have become a tenuous venture, and one that took a long time for me to recognize. However, as quickly as I understood how those triggers were impacting my life I determined to make some adjustments. I've indexed a few of those personal learnings so you have an concept of some of the things you could art work on. This isn't always an exhaustive list and a number of the ones might not test to you. I inspire you to assess your very personal scenario and in reality test what triggers you to behave in a bad manner. With that, right here is what I did:

a.   I stopped searching the news. I determined out that looking the information every day changed into allowing pretty some negativity into my existence. The ordinary barrage of political rubbish and the overall awful statistics modified into doing nothing to enhance me up. On the opposite, I found out it changed into affecting my mind-set and the conversations I might have with those round me. Aside from the occasional weather file or inventory marketplace take a look at, I avoid the records channels much like the plague.

b.   I restrained my time spent and involvement with social media. Excessive involvement with social media involvement wasted lots of my time and similar to the statistics, I modified into seeing and watching loads of terrible opinions go with the flow thru my feed every day. The negativity witnessed introduced on an further horrible attitude, which, ultimately bled into my ordinary effectiveness as a husband, father, and project supervisor.

c.   I adjusted my reaction to others emotions. Instead, I decided on to apprehend their element of view first and now not proper away react to them emotionally. I first needed to admit to myself that I even have end up with out problems stimulated with the beneficial resource of the emotions of others. Some humans do no longer like that approach as it is not clean to do. Also, revealing a weakness, now not to say running on it, isn't something many are inclined to do.

d.   Probably the most vital lesson I discovered changed into to understand that I am an introvert and information my emotional limits. As a venture supervisor you are continuously round people and for an introvert, too much human beings interaction is mentally, and emotionally laborious. At first I used "being an introvert" as an excuse to keep away from the human beings interactions and relationship building this is so crucial in a fulfillment venture manipulate Over time, I also decided out that I need to workout my "people muscle", which enabled me to spend more time spherical others without getting as worn-out. But, learning what enabled me moreover helped me to apprehend what precipitated my limitations. At my middle, I'm even though an introvert, so you might not see me socializing for hours on end like my extroverted opposite numbers. However, I actually have decided to art work thru this vital functionality of interacting with others. Without this expertise, I would have by no means stated or idea how I need to have taken my extraordinary strengths as an introvert and growth them to be a better mission manager and leader.

Making your self privy to your triggers and shortcomings will permit you to in making a few minor changes while you cope with and react for your environment.

2) Ask Others To Assess You

After you've taken the time to evaluate your self, ask others round you to assess your strengths and weaknesses. This exercise may be completed on a non-public and expert stage. However, whilst you consider that we're discussing task manipulate, discover those that you have worked with on contemporary tasks. Some may enjoy that you are looking for a few kudos or ego stroke but ensure to allow them to apprehend you are attempting to get an honest assessment of your strengths and weaknesses. Assure them that this try is so you can beautify as a undertaking supervisor and professional.

3) Complete A Formal Assessment

There are severa assessments you can take online to assess your emotional intelligence. Some are unfastened and some are paid. Pick one that works the brilliant for you so you can get an superb concept of in which you take a seat on an EQ level. You'll find a few resources for you inside the beneficial useful aid segment of the e-book.

Chapter 2: It's All About Them

A Project Manager's Social Awareness

Once you begin to understand yourself, your emotions, what makes you tick, and your triggers, then you could start to apprehend others round you.

Let's move lower decrease again and take a look at Matt's social attention for a moment. As stated, he started out out to expose frustration in his interactions with anyone. What he didn't understand become how his reactions had been affecting the group and his effectiveness with the institution. With Matt's lack of social reputation, he disregarded some essential clues that might have introduced the institution collectively and made the task run a good deal smoother. What Matt changed into lacking, inside the location of social focus, is some of what we're approximately to cowl.

Are you a reader?

Not to be wrong with reading scholastically but are you a reader of humans? High social focus comes with the capability to have a look at humans. Can you examine their faces, their body language, or recognize the variations in tone even as they'll be speaking? A undertaking manager's functionality to have a examine people and the scenario is a key differentiator a number of the average and exceptional venture managers. Let's look a bit deeper at how you may increase the skills vital to observe humans and situations. These capabilities alone will make you stand out as a mission supervisor and a pacesetter.

Reading People

There are three regions to popularity on in terms of studying human beings. One is studying their faces or their micro-expressions, 2d is analyzing their frame language, and 0.33 is listening carefully to their voice or their tone. Understanding the way to have a look at human beings can prevent quite some time and heartache for my part and professionally. Let's first test the micro-expressions.

Micro Expressions

Have you ever been in a verbal exchange with a person and discovered that they made a face while you said some thing. Most folks brush aside the ones facial expressions and maintain with our conversation. But, what in case you had been to start analyzing what those micro expressions are, what they suggest, and a way to react to them? Maybe you'll become referred to as the face whisperer. Ok, in all likelihood now not. But, you will be able to select out out up on subtleties that most in no manner will. Most of this in truth has to do with knowledge what the ones expressions suggest after which clearly looking at. According to Dr. Paul Ekman, a former psychology professor on the University of California at San Francisco and researcher on nonverbal conduct, as it pertains to facial expressions and gestures, "Whether we stay in China, Cuba, or Canada, each person specific the same seven regular feelings of anger, fear, unhappiness, disgust, surprise, contempt, and happiness." Dr. Ekman grow to be additionally the consultant to the well-known tv show, Lie To Me.

Let's observe a pair of those expressions and talk how you could observe them for your each day existence as a task supervisor.

Surprise and Anger

These feelings regularly move hand in hand because of the fact one either triggers or follows the other. For instance, if you are sharing a put off in a project, your stakeholder can be surprised, and, due to the fact you're talking a delay, this will be followed by using manner of anger. Later we'll talk why you in no way need to surprise a stakeholder. For now, emerge as aware of the micro expressions after you have added your message to your stakeholder. You will need to make certain the phrases and moves that the stakeholder is speakme are congruent with the micro expressions they may be displaying.

The reason for that may be a few can also show that the whole thing is good enough, but then cross right away to your boss or theirs and file their discontent. It's not that you'll be able to prevent them, however if you can apprehend the micro expression it offers you an possibility to engage similarly conversation with them to supply their apprehension down and display how you have got a solution or, at a minimum, show you're on pinnacle of things of your project. At the same time, you will need to anticipate what their next skip may be based totally totally totally on their expression. Also, if no in addition words are spoken, you can at least provide your superiors the heads up on what took place and the fact that the individual you brought the message to come to be no longer happy. Again, no surprises on your boss are an exceptional concept as properly.

Contempt

Why would it not be an superb concept to recognize the contempt micro expression?

Contempt is the opposite of empathy. It leans extra toward arrogance and no longer being involved what someone is going through and handiest involved with oneself. So, if you added the equal facts to your boss as above and you show a micro expression of contempt, your boss might anticipate you actually do no longer care that there is a cast off in delivery. Understanding this could will let you direct in addition dialogue to solutions and options that could advantage your boss and her worries.

One thing to apprehend, in an effort to are available in a future e-book and is a massive hassle of undertaking manage, is the cultural components of those micro-expressions. Different cultures have excellent strategies of expressing the aforementioned expressions. Reading those expressions the incorrect manner will have effects within the precise opposite manner which you supposed.

We've best scratched the floor and this is in no manner an exhausted listing of micro-expressions, however in case you take some time to test this type of facial recognition, you may find out approaches to speak and anticipate extra correctly.

Body Language

Body language, on the equal time as now not continuously so subtle due to the fact the micro expressions you may see on a person's face, can give you an component at the identical time as speaking to a person and help collectively together with your relatability to that person. Let's test what Matt taught me approximately body language.

After several months on the venture, you may see Matt's frame language exchange. He emerge as more slumped over, shoulders rolled forward, and his head was more down than it changed into up. I ought to tell some difficulty grow to be wrong, however I had no longer however developed the talents to apprehend what I became looking at. However, fortuitously for Matt, a person inside the room did. They want to have spoken to him because of the reality every week later, in our subsequent meeting, he changed into sitting up right away, his voice have become greater forceful, and you could inform he had command of the room. I turn out to be very curious about what had changed, so I requested him and this is what he counseled me.

Matt endorsed me that a person had pointed out how his body language grow to be being perceived and the manner that contributed to how he come to be acting. He went on to tell me that he no idea he become doing this but emerge as feeling increasingly more pressure because the undertaking went on. I even have come to be however curious so I requested him to inform me greater approximately what he found out.

Confidence

One of the primary topics a task supervisor should deliver is self assurance. Without self guarantee, you may have a difficult time getting group individuals to do what you need them to do. You can discover ways to examine one in every of a kind's self guarantee stages, however the first character you need to have a have a look at is you. Are you portraying self belief as you take a seat down, stand, and stroll?

I referred to earlier than the way it changed into stated to Matt that he grow to be slumped over at the same time as he sat. It come to be moreover noted to him that his legs have been crossed at his ankles and his arms have been often crossed. This form of frame function tells people, whether or now not they recognise it or no longer, that you are in a protecting or closed off feature. When you are in this kind of feature you are not portraying self guarantee. I will encompass a link in the assets segment so that you can see exactly what this seems like. Let me display you a manner to do this within minutes.

There are a few subjects you can do to right now increase yourself assure and the way your self guarantee is regarded via others. This is based totally mostly on scientific studies from Amy Cuddy, a Ph.D. Professor at Harvard Business School. Let's examine Matt's frame characteristic. I said his hands and ankles have been crossed. So if you have been to teach Matt the first hassle you may inform him to do is uncross his fingers and ankles.

By uncrossing your limbs you open yourself up. When you're more open, you are taken into consideration as greater trusting. According to Cuddy, on the same time as people meet someone for the number one time they'll be comparing subjects. The first is, are they sincere. The 2nd is, are they assured or succesful. As I said, through beginning yourself up you robotically offers a layer of trustworthiness in evaluation to a person who has a frame feature this is closed off. Being open additionally units off signals that you are assured. This is due to the fact a person who is confident would not need to be hiding in the again of anything or shielding themselves from sincerely all people. After all, they may be confident in who they'll be and what they may be doing.

So if you were in a meeting with Matt, and you noticed him sitting in one of the positions I've stated, which one could you take delivery of as real with extra and assume is more confident or organized? Think approximately someone you've been round or have worked with a similar posture. Do they stumble upon as honest or confident? Are they able to command interest or have an impact on the ones spherical them? Why is this important to you as a venture manager?

As we previously pointed out, yourself-awareness is an vital key to growing your EQ. However, being aware about others' frame language may additionally moreover provide you with some notion into specific people you discern with and apprehend how powerful they may be as a member of your organisation. Not excellent that, but this can offer a route which will apprehend a manner to have interaction with every body. Let's have a look at an instance.

Let's say I am going for walks with an engineer on a task and I see that he's sitting much like Matt, with arms and legs crossed. Also, I may moreover have a have a look at that he regularly doesn't look up or into my eyes even as he speaks. If you have been to begin speaking to him like someone with high self perception, you could cause him to shut down. For example, if there can be a undertaking this is had to be completed and also you technique this person aggressively, which a assured character would probable have little issue with because of the reality they may most likely respond to you with comparable aggression, they may do what you need them to do however within the method they'll lose understand for you. While some project managers won't to begin with care approximately dropping recognize, as long as what they need receives finished, this can impede destiny improvement because the project and employer build momentum. As a assignment supervisor, the final trouble you need to do is sluggish your project institution's momentum. So how might also want to you approach someone like this? Let's talk thru one way to do that.

For a person who lacks self notion or is extra closed off we will make a few assumptions. One assumption is that they're now not comfortable speakme with others. You will discover diverse degrees of this in the place of job, however the handiest manner to permit someone to open up is find out what their pursuits are. I don't suggest their work or career hobbies, however their private hobbies. Yes, that's right, you are going to should get private. You can do this through in fact taking a look at their table and what they pick out to expose. What's at their desk is a hallmark of what they may be open to speaking approximately.

Chapter 3: Self-Management

So what's strength of will and why is it important to us as project managers? As a task supervisor, you're held accountable and accountable for how a task runs and its outcomes. In essence, you are the best throat to choke in terms of the project. While it's miles actual that the venture supervisor isn't always the only virtually liable for the fulfillment or failure of a project, stakeholders will almost commonly look to you at the same time as topics are going wrong. Typically you could not be doing any of the programming or constructing of the product but your stakeholders will appearance to you to answer for and be accountable for the first-class and the lousy that takes region for the duration of all degrees of the assignment. Because of this, there are frequently some of demanding conditions that get up in the direction of a venture. How you manage those situations and, greater importantly, the way you manipulate yourself could make the distinction the various achievement and failure of the mission and whether or not you're provided greater tasks in the future.

Where do you begin when it comes to willpower? We previously stated triggers. Once you understand what your triggers are you should discover ways to react to those triggers. This is in which self-management is available in. The first element to do is begin to maintain a mag. You can't control some factor you're now not aware about. Taking the time to listing out your triggers might also additionally additionally look like a waste of time to a few, however as soon as what makes you, harassed, angry or excited, you could manage those feelings extra effectively.

In its excellent form, you can see a low EQ and espresso strength of mind while watching a toddler throw a tantrum. Can you don't forget you studied of a time on the identical time as you had been running with or round a person who reacted like a infant throwing a tantrum? I can will let you understand I in truth have and I'll a share a totally particular tale with you.

During taken into consideration one among my initiatives there was a technical problem that took place and one of the engineers on my undertaking, Charlie, become not capable of restore the problem and had to do some extra troubleshooting. This hassle have emerge as at the computer of an authorities. Charlie came to me with a look a worry on his face and stated the purchaser wanted to look me. I knew this could not be well, given the advent on Charlie's face when he got here in. So Charlie and I were given inside the elevator and head as a lot because the twenty first floor. This government had a gaggle of initials after his name, which include VP and MD so he changed into very important, or so he idea he have come to be. Unfortunately, for us, he have end up moreover seemed for his loss of people abilties. As we entered his place of job, I greeted him and asked him what the hassle grow to be and how I can also want to assist solve it right now. He proceeded to tell me how he grow to be late for his next assembly and could not get a file to artwork well. The greater he spoke and described his hassle the extra red his face have come to be and the louder his voice have been given. After several minutes he began out pounding his fists at the table, yelling at us that this needed to be constant now and made numerous threats for added degree. As if that wasn't terrible sufficient, he abruptly stood up sending his chair flying into the wall within the back of him growing a noisy bang, as he stomped out of the room to wait his next very vital assembly. Stunned, Charlie and I in reality sat there looking at every one-of-a-kind in disbelief. Just then his admin, an older lady with a completely calming voice and appearance on her face, walked into the room and stated, "don't worry boys he's now not mad at you." Ha, you can have fooled us, I idea.

Luckily for Charlie and I, the whole state of affairs occurred so speedy that neither humans had time to react. A reaction, of any kind, also can have had a no longer so favorable final results. So why could probably someone at one of these high level with obviously a high diploma of schooling, given the extensive sort of initials after his name, act that way? He became an clever and nicely-knowledgeable man or woman, but he definitely displayed a totally low EQ in our brief come across with him.

The bottom line is he did now not recognise the manner to self-discipline and modified into virtually now not aware about his triggers. If he changed into self-aware, he selected to push aside his triggers. Now you cannot enjoy this sort of interplay all through your profession, and quite frankly I wish you never do, but you'll be tested. There are going to be the ones people that push your buttons. Your achievement will not come from just dealing with them however greater of it'll come from the manner you manipulate your self even as you're handling the ones traumatic conditions. This brings me to my subsequent element with self-control and that is willpower. Self-Control

As I referred to, you will be examined. Having the strength of will to no longer react the manner you experience is essential. You can be questioning that is pretty obvious, but don't forget our VP pal Charlie and I encountered. You have to suppose that someone together alongside with his pedigree and feature could probably have his stuff together, however he did now not. You need to bear in mind that inside the paintings surroundings people are underneath various quantities of pressure. Depending at the quantity of pressure and their level EQ, will rely upon how they react. The essential aspect to keep in mind is that it isn't always about how they react however approximately the way you react. You can quality control man or woman and this is you. Let's check some unique strategies you could increase your self-control.

Look on the huge image

As task managers, it is able to be very clean for us to get into the weeds, and maximum human beings do. While it can be critical to dive down into the minutia, it is not a place wherein we are presupposed to live long time. You will want to take a look at and apprehend the larger photo. Ask yourself, what element does this situation play in the big picture? Is this something that would have an effect on the very last results of the challenge or do you really need to address the scenario and pass on? Clearly knowledge this could permit you to have a better angle and will allow you to manipulate the state of affairs and your self.

Relax and Stop

When you are within the throes of a large task and the tension is mounting it is crucial to loosen up and simply forestall to take a breath. The cause this is so essential is that the act of stopping will help you avoid making impulse choices. Too often humans in immoderate-pressure situations react or make alternatives all of a sudden. They see and experience the need to show a sense of urgency and dive headlong right right into a selection that would have catastrophic consequences on the general venture. However, in case you take some time to loosen up and count on you decrease the possibilities you could overreact. So, how do you do this?

One of the most effective methods to lease a manner called box breathing. To try this you'll need to find out a quiet area wherein you may with out issues sit down down. Next, near your eyes and breath in for four seconds. From right proper right here you'll need to maintain your breath for four seconds. Then, you will exhale for four seconds. Finally, you may maintain your breath for some other 4 seconds. Once you've completed this very last breath hold you may start the manner over with the useful resource of breathing in for four seconds and so forth. Do this as frequently as it takes a good way to begin to revel in greater comfortable. When you first start this exercise you could experience barely dizzy. This may be due to the truth you are not used to taking the time get that a bargain oxygen into your device. I endorse you do that only some instances at the start to begin so that you can exercising and be aware about how your frame is reacting. It is likewise crucial to breathe through your stomach in desire on your chest. This will allow for deeper breathing and a greater calming effect.

Why does this paintings? Without getting too scientific, this kind of respiratory turns on your parasympathetic nervous gadget, which enables your body manipulate things like digestion and lowers your coronary heart rate, for instance. These subjects arise even as you are in a more relaxed u . S . And now not making ready for combat or flight

Sleep

A lot of assignment managers spend countless hours, and lots of an overnighter, taking walks and deploying their responsibilities. However, without excellent enough sleep, your desire-making and self-discipline will undergo. I do need to ensure you understand the element sleep performs on your potential to manipulate yourself and your emotions. Look once more at the toddler example. If you have got ever been round a infant at the same time as they are drawing close to their bedtime you can see a candy and loving child turn into a hint demon, nearly as if looking a werewolf redesign, as the overall moon seems. You don't want to be that person, so make sure you get sufficient sleep.

Self-Discipline

Another issue of self-management is electricity of mind. Whether you're new assignment control or have been coping with tasks for numerous years, you recognize how easy it's miles to get distracted. At art work, you have humans preventing through your table, smartphone ringing, electronic mail, without delay messenger, and so forth. This doesn't encompass the quantity private distractions like texts and social media. The quantity of distractions is limitless. One of the subjects you need to get in fact well at is self-control. As professionals, we understand this, however understanding and setting it into movement are absolutely various things. Since maximum tasks have so many variables occurring it's far clean to get distracted with the problem of the day. Let's examine some strategies you could enhance your strength of will.

Start Small

When seeking to alternate a addiction it's miles often incredible to transport for the low placing fruit. You recognize the ones matters which may be exceptional to deal with? I don't understand about you however I get masses of emails each day. Some are important for me to react and reply to right away however most of them I am simply copied on so I am stored in the loop. This become a supply of distraction for me and one I had to discover ways to manage. Now I will admit, I am in spite of the reality that a bit in development with reference to e mail however I'm certainly tons better than I become as soon as. Several years ago, I determined myself spending hours each day analyzing, considering, and responding to emails. So loads in order that it become beginning to have an effect on my paintings output and cognizance. So I determined to make some modifications and I can let you know the benefits have given me once more numerous hours in some unspecified time in the future of my week. What I did modified into keep my e-mail off and excellent checked it in some unspecified time inside the destiny of advantageous instances of the day. This way I have become controlling the float of information instead of reacting whilst a few factor came in. I'm not dealing with existence and death conditions so I did no longer need to fear about being on name every minute. This freed me as a whole lot as do the matters I want to do to make development on my challenge. The caveat to this e-mail method is it did take me some time to recover from the addiction of getting to test e mail all of the time. Email, for some of people, gives them a experience of significance and something to occupy their time. The truth is only some people actually need to be to be had for each e-mail this is available in on a 2nd with the resource of way of second foundation. Over time, I actually have emerge as capable of create the energy of will to test my e mail in a few unspecified time inside the destiny of positive instances of the day. Even despite the fact that this adjustment grow to be very small it paid off in large dividends even as it got here to my time. So pick out some issue you can you could modify that is small and plausible in advance than you start tackling the huge stuff.

Get Focused

We've already pointed out the sort of distractions we come upon at some point of our day, so getting and staying centered is quite critical. However, I'm now not talking approximately being centered for hours at a time. Quite frankly, only a few can try this. What I'm talking approximately is a short burst of strive the usage of what is called the Pomodoro Technique. According to Wikipedia, "the Pomodoro Technique is a time manipulate method advanced via Francesco Cirillo within the past due 1980s. The technique makes use of a timer to break down art work into intervals, historically 25 minutes in period, separated through quick breaks. These intervals are named Pomodoros." There are a number of timers you may download out of your app save or with the aid of looking on Google. You may even simply set a timer on your smartphone for 25 minutes and use that. I'll ensure to encompass resources on this problem count wide variety.

One of the crucial element factors of this technique is to take away all distraction and excellent popularity at the venture accessible for 25 mins. Shutting down your email for this quick time will help. Do you believe you studied you and your international can live on without you checking email for 25 mins? This focused hobby does essential matters:

1) The most obvious is it leaves you freed from distraction so you can get extra finished ultimately of these 25 mins

2) even as you offer yourself most effective 25 minutes to finish something the experience of urgency is advanced. This allows you to faucet into your creativity and bring greater of the excellent out of you because of the reality you satisfactory have a short time frame in which to perform your task.

So if you need to be extra self-disciplined, get extra focused in quick bursts.

Fail

Okay, this one also can appear counter-intuitive. After all, the last problem you need to do as a challenge manager is fail. What do I imply by means of manner of this? I propose do no longer worry failure. With the Pomodoro Technique, I described, there are some that might not even attempt due to the reality they may be afraid they'll fail to complete the project within the allotted time. So in desire to taking the danger, they in no way begin.

A commonplace trait I've found amongst many task managers, and maximum people in popular, is their adversity to failure. Most view failure as a bad aspect at the same time as in truth it's miles no longer. As a keep in mind of fact, failure is crucial to success. Why is this? The smooth cause is at the same time as you fail you are commonly on the out edge of your capability set or comfort place. Identifying those edges allows you to analyze extra and broaden.

If you're capable of view failure as an opportunity to decorate you could end that the greater and faster you fail the greater and faster you could decorate. Fear of failure is often visible among project managers whilst reporting recognition. Many are afraid to record a failure, a omit, or a shortcoming on a project due to the truth they worry it'll probable be a reflected image on them. I used to try this myself. I'd try to paint a rosy picture on the the the the the front forestall even as I frantically labored on the once more give up to repair the trouble so no one might understand.

What I didn't recognize become that now not reporting the pass over, issue, or risk have grow to be an opportunity to talk to those that would probable assist. In my want to not fail or appearance incompetent, I changed into missing out on creating alliances with the very human beings that might make my existence much less hard and help make the task a success. So if you need greater power of will learn how to now not be fearful of failure.

You understand yourself higher than each person, so you can likely offer you with severa greater methods to be greater self-disciplined. I inspire you to take some time to extend this thing of willpower and you could see the extremely good outcomes in a brief term.

Self-Talk

This difficulty rely may additionally moreover sound everyday to you in terms of assignment management however the truth is that what we're pronouncing to ourselves on a each day foundation has a profound effect on how we carry out. This isn't some woo-woo pseudo-scientific nonsense. Your self-communicate creates bodily and neurological adjustments in your thoughts. For example, if you have had awful self-talk for numerous years then you definately definately are reinforcing the grooves in your mind that useful resource this sort of questioning. The same goes with the notable mind. When you attempt to trade your self-talk you're growing new styles and connections to your brain. Consequently, you're developing new pathways, similar to cars the use of down a contemporary dust avenue. Over time, grooves may be shaped on this road with each passing vehicle.

Think lower decrease returned to the final come across you had with a person. It could have been a hallway communique, a standing assembly, or maybe handiest a meet with buddies out of doors of hard work. I will wager that prior to or at some point of that communication you have been announcing things to yourself. Let me offer you with an example.

Early in my profession, I changed into given the opportunity to take over a assignment that wasn't going so well. I grow to be new to task manipulate and the man or woman I have become taking over for emerge as a whole lot greater skilled than I grow to be. I had all forms of doubts and thoughts going for walks thru my head as I commenced to art work in this mission. I endure in thoughts questioning such things as; I don't have sufficient enjoy and what if I fail? Is every body going to be aware of me? What if I screw up? What if I disappoint my boss? The list went on and on. Have you had comparable thoughts?

42

What I didn't recognize then changed into that the ones thoughts, even as a few being very legitimate, have been not helping me pass ahead. This self-talk emerge as in no way going to get me to a place in which I may additionally want to perform at my superb. If I had been to move lower lower returned to younger Bryan and deliver him a few recommendation on this vicinity I may additionally say 3 subjects:

1) Become self -conscious. Sound familiar? The first problem I'd do is to become aware of my self-speak. What am I pronouncing to myself? We all speak to ourselves so it is not such as you need to pass an extended way or dig deep to honestly be conscious. Like we've got talked about, being aware of what you're doing and saying is a part of your emotional intelligence. With this primary aspect, you're really searching at. There is not any judgment. Your self-communicate sincerely is what it is and not whatever extra. Your undertaking in this step is to perceive what you're pronouncing to yourself and write it down.

2) Modify any terrible self-talk to three factor more excellent and in a manner to be able to flow into you ahead. For example, in case you say to yourself, "I'm not a remarkable public speaker" then trade this to "When I speak human beings are inquisitive about what I want to pay attention". Does this imply that after you begin saying this that human beings will need to pay hobby what you've got have been given to mention? No, they received't. However, over time your belief will purpose you to behave in a distinct way. This exchange is what is going to cause humans to be interested by what you have got to say. Your body language will alternate, your voice tone, inflection, and quantity will exchange. All the components that would make you someone humans want to take note of you'll start come thru. The reason for that is our unconscious mind is type of a warm temperature-attempting to find missile. It can not tell between truth and imaginary. If you inform your mind a few issue it will start to art work to show that right. The rational a part of our brain that tells us what's actual or imagined is

44

our conscious mind. Don't do not forget me? Think back to a film that had a unhappy or frightening factor in it. Did you cry or have been you scared? Was what changed into taking location in the film actual? Of route now not, but you still reacted. If the film have become sad you could have cried. Was there any logical reason an amazing manner to cry? Your unconscious mind grow to be reacting to the imagined state of affairs supplied within the movie. Likewise, we are able to start to create our very private truth by way of way of changing what we're saying to ourselves.

3) The final step is repetition. Your idea techniques and habits have come from years and years of self-talk. It will take time for those new pathways and connections to make bigger. I've take a look at that it takes 21 days to create a dependancy. However, after doing plenty of studies, I determined it takes a hint bit longer than that. If you will make a protracted-lasting change I may additionally want to suggest repeating some factor self-speak you pick out for 60 days. The longer you

repeat your message the deeper the grooves on your thoughts turns into and the longer the idea system will final. Eventually, repeated sufficient instances, the new concept sample becomes how you mechanically think. Depending on in that you're, your historical past, and your modern-day-day self-talk it may take greater than 60 days. My recommendation is you do it long sufficient till your self-speak has changed with out you having to consciously do not forget it. Just begin with one message at the start. Remember, begin small. Now which you've started to enhance yourself-speak it's time to test some different trouble of self-management and that is your conduct.

Habits

For this part of the e-book, I'd want to percent an excerpt from my first e-book on Habits. You will see that your behavior tie right away into what we noted approximately power of thoughts.

How often do we understand what's nicely for us, but we fail to act? The location of acting on what we apprehend to be genuine, in location of what we experience, allows deal with the uncertainties of existence. Growing up, it appeared that most of my options had been rooted in reacting inappropriately to my feelings—doing what I felt rather than doing what I knew I want to do.

Chapter 4: Relationship Management

As a project supervisor, your venture will stay and die by way of the use of using the relationships you are able to create and preserve. In this section, we will discuss a few strategies you can put into effect to now not handiest increase your EQ but solidify your relationships and make certain the success of your mission.

Be Interested

A lot of assignment managers I've are to be had in contact with spend more time being interested by what they want for themselves. Unfortunately, what they want isn't always commonly what will make the undertaking float ahead or affect others to do what the challenge supervisor dreams. If you need to assemble relationships collectively together with your stakeholders and your corporations you have to first be interested by them. Let's first begin with your stakeholders.

Your stakeholders are the ones paying on your project so that you can now not take the time to really apprehend what they need is like getting ready to supply some difficulty they do not need or need. Understanding what they want starts with first expertise them; their role, their temperament, and possibly a number of their extended-term desires for the task and their very private profession. If you have not worked with the ones stakeholders the primary issue to do is to begin to bring together consider. The quality manner they will consider you, with out you having introduced some issue, is within the occasion that they get to recognize you and prefer you. Until you could display you are able to deliver, this is all they have to bypass on. So in case you're interested in building accept as authentic with and building a dating along with your stakeholders you need to first be interested in them.

Building a relationship together with your team may be very similar to your stakeholder, in that you want to build accept as true with with them. Because of your feature due to the fact the challenge supervisor, there may be a positive quantity of blanketed believe, but, you can have the ones that don't don't forget all people till they show themselves. In both case, like collectively at the side of your stakeholders, you have to first get to recognise them and they need to develop to love you if you are going to get them to really do not forget you. Can you correctly entire a task with out your institution liking and trusting you? Sure you could. However, your life may be a remarkable deal a whole lot less complicated in the event that they determine on and undergo in mind you as you will have more have an effect on over them. In his ebook, Influence: The Psychology of Persuasion, one in every of Robert Cialdini's pillars of have an effect on is likability. He states, "Few humans might be amazed to examine, that more frequently than no longer, we most choose to say fantastic to the requests of someone we recognize and prefer. What is probably startling to be aware, however, is that this easy rule is utilized in loads of techniques thru total strangers to get us to conform with their requests." If this is real, and I take delivery of as genuine with it's far, how lots greater are you able to get executed and how many extra human beings can you have an impact on in case you enhance your likability?

Depending on the enterprise you are in, you can now not have devoted sources. In distinct words, the assets assigned in your assignment might also have responsibilities on distinct tasks, which suggest which you could not typically be their top priority. One of the keys on your success as a task manager is to make your venture the number one priority within the minds of your institution. One of the amazing techniques to do that is to create a dating with them and increase that relationship so that they decide upon and get hold of as real with you more than one of a kind challenge managers, whose time you may be competing with. If you are lucky sufficient to have committed assets operating to your venture, then loads the higher. However, if you are going to get the maximum out of your group, construct sturdy relationships with the useful resource of being inquisitive about them.

Care About Your Team

One of my desired costs is with the useful useful resource of Theodore Roosevelt and he stated, "People don't care how hundreds you realize till they understand how loads you care." Through the years many human beings have quoted the ones phrases but the truth remains that during case you want people to care how plenty you apprehend you're going to need to care approximately them first.

In our international of cut-off dates, instant gratification, and quick hobby spans, maximum are used to simply shifting via their day at a quick and worrying tempo. Especially for those those who're assignment managers, we have to deliver on time and on charge range, so we want to strength, strain, force. Right? Well, positive we do, but it does not endorse we can't make an effort to care approximately our group.

I surely have labored with some extraordinary challenge managers whose care meter is off the charts. It's the little subjects. For instance, for every early morning assembly they have got, they create donuts. During crew conferences, they in my view apprehend the efforts of those going above and beyond. If a few issue goes on, in my view, with a member in their team they make the effort to ask and display proper trouble. All the ones actions are not finished out of looking to govern all of us. They do those gadgets due to the reality it is who they may be at their center.

I will in no way forget about a challenge I became running on, no longer as a venture manager, but as a technical useful resource. We had had an extended night time time of installation and troubleshooting and the group became worn-out. The subsequent day, I had a card on my table with a have a observe that take a look at, "Thank you for doing more than what is expected." Nobody else ever stated something about the paintings that grow to be completed however that clean gesture made the hours and attempt surely well worth it. I will permit you to realize that each time this challenge manager wanted some detail I changed into willing and prepared. Feeling cared for is a fundamental human want. Give it freely and it'll pay dividends almost about building relationships.

Be Empathetic

Empathy has generally been thrilling to me. At our center, plenty of us just need to be understood. As teenagers, we go through this because of the truth we don't suppose our mother and father or the arena is aware us. When we get into the staff empathy has a way of getting misplaced. It's those which could show empathy that stands proud.

Understanding someone's emotions takes effort and time to apprehend. Remember I stated how you can have assets wherein your project isn't always the great undertaking they're jogging on? Understanding their state of affairs once they have or extra project managers vying for their interest can bypass a protracted way. One issue that I've visible art work nicely is taking a number of the burdens from them by means of manner of talking to the opportunity challenge manager and working thru any scheduling conflicts. First, but, ask the aid if that is some thing they can be precise enough with you doing on their behalf. Most of the time you are not their direct manager so some can also additionally take offense to your handling their time outside of your challenge. However, in case you communicate to them first and display them which you apprehend the state of affairs they're in, they may be more than inclined to have you ever ever artwork subjects out for them. This can circulate a long way to building a strong relationship with that individual.

Stakeholder Management

We've taken a while to speak about courting manage because it relates to your organization. Now permit's test your stakeholders and how you may construct and manage the ones relationships. According to pmi.Org, "Successful initiatives depend on a number of human beings, and it's far the clever assignment supervisor who actively determines who they may be and what regions of the challenge they have got an impact on. A forgotten stakeholder often rears his or her head on the maximum inopportune time, wreaking all forms of havoc within the assignment. But many assignment groups do a negative project of figuring out project stakeholders and gaining their self-control to the mission."

Stakeholder manage will employ a number of the skills we mentioned earlier but it's going to utilize numerous extra capabilities a good way to gain you in future control roles, need to that be a direction you decide to pursue. One such talent and probably one of the maximum critical to any assignment supervisor and leader is verbal exchange. Let's take a look at how to make use of powerful verbal exchange to collect stronger relationships.

Communication

Although the artwork and ability of powerful communique are so critical to being an powerful venture supervisor and in dating manage, it is by the usage of far one of the most tough abilities to realise. Why do you discovered that is so difficult?

There are probable loads of motives why, but I will offer you with my opinion on this project. My opinion has been normal over the last two decades from, pretty frankly, messing up A LOT in this region. To within the meanwhile I'm nonetheless mastering and will hold to analyze with every mission I work on. So what do I count on the reason communication is so difficult? I receive as genuine with it's miles due to the truth each person spend maximum of our time considering ourselves. Yes, I recognize that may be a ambitious assertion to say sincerely all of us. But, bear in mind this for a minute.

For the majority of the day, are you consider other humans or are you considering the stuff you want to do or getting misplaced to your private thoughts? You are considering you, proper?

There isn't always anything incorrect with this, in fact, this is how we're stressed out. From the time you've got been a little one you have been targeted in your desires; being fed, modified, and cared for. As you grew up, you started out to take care of the ones needs however the desires remained the identical. Throughout your day, you're planning what wants to be achieved or doing belongings you need to do. As a venture supervisor, you're centered on getting the mission completed as a part of your approach. As you try this, you're considering what you want to do. Again, none of this is inaccurate, it definitely is. Have you ever been in a communique in which you were on the lookout for to think about the subsequent element to mention to the person you have been speaking to, instead of truly paying attention to what they have been pronouncing? I realise I surely have, normally.

In my case, due to the fact I even have turn out to be targeted on what I become going to say subsequent I wasn't sincerely listening. My interest and interest have been on my next incredible announcement. If you were to ask a group of adults what they perception one of the maximum essential factors of verbal exchange is, I suppose most may agree that listening is probably at or close to the pinnacle of the listing. If you'll be sturdy at dating control you'll should become very sturdy in the artwork and ability of communication. The first step to being a sturdy communicator is to end up a brilliant listener.

Listening is not absolutely listening to what they'll be saying, however it's also looking at what they do now not say. Take the situation I will describe in the international members of the family segment. You will observe that Tom, a supplier I became engaged to artwork with, have become telling me he preferred to ensure he have become brought in and contacted first earlier than every body came on internet page. As I listened to his voice, his tone and frustration said masses extra than that. He turned into searching out to be supported, heard, and revered. When it includes listening, what humans don't say is form of greater critical than what they do say. Take the time to decorate your listening abilties via way of the use of some the strategies below.

Do:

- Maintain eye contact

- Be present

- Ask questions to benefit information

- Provide comments

- Be open minded

Don't:

- Look at your smartphone or pc

- Interrupt

So why are there greater Do's than don'ts at the list above? The reality is that for each Don't there are various Do's. I need you to reputation more at the Do's as you increase your listening abilties. This isn't always an exhaustive list, but you can find out many more assets obtainable at the way to be an effective listener. I advise you make the effort to actively look at this issue count number.

Since communique is this form of key detail to building and keeping relationships, allow's examine a few more factors of communique past listening.

No Surprises

Since communique is critical, it's miles critical to speak each the coolest and that awful information. In every challenge you determine on, there can be a few element as a manner to get up this is unexpected. In the case in that you want to deliver statistics about a put off or alternate, it's miles higher to provide your stakeholders early phrase. While they may be dissatisfied, letting them realise early is a lot higher than having to inform them after the remove has occurred. Worse but, they find out thru other assets. Let's check an instance that Matt treated.

It end up round mid-12 months and Matt observed out that, through one in every of his carriers, that over 1/2 of the device he have end up looking ahead to to be introduced emerge as going to be not on time thru way of severa months. This changed into something that emerge as genuinely out of Matt's manage and he had no possibility but to alter the modern-day time desk. His planned circulate-stay date became the forestall of the 12 months, but after adjusting the time desk because of the delays, this have become going to push out the pass-live date to the stop of the first place of the following year.

As quickly as he had the modern dates, he walked over to his stakeholder's office to allow them to understand the news. While they had been not glad to pay attention about the postpone and at a loss for words Matt on other alternatives and possible mitigation, they understand that Matt had finished all he must to supply the answer as rapid as viable. Matt made superb to document their communicate then allow his enterprise and every different occasions that needed to understand, approximately the postpone.

In this situation, at the equal time as pretty simplistic, Matt did not postpone or hesitate in letting his stakeholder understand what turn out to be taking place as fast as viable. By doing this he did important matters. 1) He did not allow his stakeholder to be surprised with the aid of the "lousy information" 2) By letting the stakeholder realize early he gave his stakeholder the possibility to assist come up with an answer even as there was nevertheless time. One of my largest learnings early on in my career became a manner to attend to the delivery of terrible facts. Unfortunately, I located out this the hard way by way of doing the exact opposite of what Matt did, and that is to attend to talk. What I did as an alternative have become try to repair the problem on my own, in a vacuum, and not on time telling my stakeholders what come to be taking place. I did this because I modified into terrified of their response or that I may probable get in problem. The fact is, as a mission supervisor, it is your interest to file the notable, the awful, and the ugly – and as quickly as viable. You are not out to win a reputation contest. Where I end up afraid of getting in trouble for delivering some now not so favorable data, I modified into really risking a greater wrath with the useful resource of delaying the inevitable.

Communicating regularly and early is vital to the success of your challenge. While you'll be the bearer of lousy data at instances, you are the lead communicator to your crew. So, take the important steps and make certain all who need to understand, realise early and often.

Diplomacy

According to dictionary.Com, the definition of worldwide family members is the "ability in dealing with negotiations, handling human beings, and so on., so there may be little or no sick will." Let me percent a brief story to illustrate this definition.

Recently I received an email from a dealer companion which without a doubt stated, "Please call me". Remember, I referred to Tom? I certainly failed to count on anything of it as we're commonly speakme about fantastic challenge related gadgets. When we had been given at the mobile phone he began out to tell me about some one of a kind provider who took it upon themselves to return to the location, in which he have become the net web web page manager, and started searching round so they may get a few measurements for a deployment we had been operating on. Well the internet site on line manager, Tom, did now not understand someone drawing close to-web internet web page with out be aware.

As we spoke, Tom's voice began out to enhance as he relayed the story to me. He stated that he is in price of the internet website on line and that if every body desires to see a few difficulty they want to undergo him. The extra he talked the extra charged up he had been given. At this component, all I need to do grow to be pay attention. He had a legitimate thing but he preferred me to move talk to the possibility company and allow them to realize the guidelines. Because I desired every to artwork collectively I needed to act speedy and make sure Tom were given what he wanted, which became to be confirmed and revel in supported. I moreover had to make certain that my different vendor understood the floor policies of the engagement at that precise internet web page.

I referred to as the offending supplier and defined the rules of engagement at that web internet web page. In doing so, I needed to make sure I maintained a awesome courting with him thinking about that we were coming close to the final date of the essential milestones and any delays or hiccups should reason the project to transport past our anticipated schedule. Although a minor incident, if not treated nicely, a scenario like this can have escalated to a point where it affected no longer handiest the operating dating of the groups but must have posed a severe hazard to the general time table and venture completion.

If you need to be effective as a undertaking supervisor you could need to include international relations for your not unusual schooling on emotional intelligence.

Stakeholder Management

We've taken a while to speak about courting manage because it relates to your organization. Now permit's test your stakeholders and how you may construct and manage the ones relationships. According to pmi.Org, "Successful initiatives depend on a number of human beings, and it's far the clever assignment supervisor who actively determines who they may be and what regions of the challenge they have got an impact on. A forgotten stakeholder often rears his or her head on the maximum inopportune time, wreaking all forms of havoc within the assignment. But many assignment groups do a negative project of figuring out project stakeholders and gaining their self-control to the mission."

Stakeholder manage will employ a number of the skills we mentioned earlier but it's going to utilize numerous extra capabilities a good way to gain you in future control roles, need to that be a direction you decide to pursue. One such talent and probably one of the maximum critical to any assignment supervisor and leader is verbal exchange. Let's take a look at how to make use of powerful verbal exchange to collect stronger relationships.

Communication

Although the artwork and ability of powerful communique are so critical to being an powerful venture supervisor and in dating manage, it is by the usage of far one of the most tough abilities to realise. Why do you discovered that is so difficult?

There are probable loads of motives why, but I will offer you with my opinion on this project. My opinion has been normal over the last two decades from, pretty frankly, messing up A LOT in this region. To within the meanwhile I'm nonetheless mastering and will hold to analyze with every mission I work on. So what do I count on the reason communication is so difficult? I receive as genuine with it's miles due to the truth each person spend maximum of our time considering ourselves. Yes, I recognize that may be a ambitious assertion to say sincerely all of us. But, bear in mind this for a minute. For the majority of the day, are you consider other humans or are you considering the stuff you want to do or getting misplaced to your private thoughts? You are considering you, proper?

There isn't always anything incorrect with this, in fact, this is how we're stressed out. From the time you've got been a little one you have been targeted in your desires; being fed, modified, and cared for. As you grew up, you started out to take care of the ones needs however the desires remained the identical. Throughout your day, you're planning what wants to be achieved or doing belongings you need to do. As a venture supervisor, you're centered on getting the mission completed as a part of your approach. As you try this, you're considering what you want to do. Again, none of this is inaccurate, it definitely is. Have you ever been in a communique in which you were on the lookout for to think about the subsequent element to mention to the person you have been speaking to, instead of truly paying attention to what they have been pronouncing? I realise I surely have, normally.

In my case, due to the fact I even have turn out to be targeted on what I become going to say subsequent I wasn't sincerely listening. My interest and interest have been on my next incredible announcement. If you were to ask a group of adults what they perception one of the maximum essential factors of verbal exchange is, I suppose most may agree that listening is probably at or close to the pinnacle of the listing. If you'll be sturdy at dating control you'll should become very sturdy in the artwork and ability of communication. The first step to being a sturdy communicator is to end up a brilliant listener.

Chapter 5: Overview

Are you a successful Project Manager? Everyone intends on being a hit on the onset of a undertaking, however it's not unusual that even a mission that come to be deliberate out to perfection starts offevolved to extend troubles due to the fact the project progresses. The purpose task managers fail with the desires once they start out inside the beginning of a assignment has greater to do with how they reply to problems that rise up for the duration of the assignment than every one of a kind purpose. Does it propose that they may be a terrible undertaking manager? No, but how they manipulate the issues speaks volumes approximately their expertise.

So what are the critical issue dispositions to be a successful Project Manager? This ebook is meant to be a guide on what I surely have found to be the key tendencies for someone who's successfully coping with a project. In every phase I will speak a trait that I determined is crucial to fulfillment and then provide you with one or extra examples of every trait so you will have a essential knowledge of the idea. These examples may be easy to understand and based totally mostly on real life instances.

Chapter 6: Project Management

Every employer is concerned in tasks on a day by day basis and could at one time or some other be concerned in a larger project. Whether it's far a few factor as small as organizing a assembly for a state-of-the-art client or as massive as constructing a modern-day skyscraper, what all initiatives have in common is that they will need a person to prepare and oversee them. Every worker is constantly worried in task management on a every day basis. Most agencies will not require a committed task manager. The want for a committed project manager comes into play while there are non-prevent obligations or the task is so massive that you want to have someone that should organize all the out of doors resources together with the enterprise enterprise sources at the manner to efficiently whole the intention that you are trying to gather.

Once an commercial enterprise employer has determined that they would like to cope with a venture, they may need to have someone manage the project, to make certain that the venture is a fulfillment. This supervisor could be the project manager. The undertaking manager is answerable for the transport of the cease product or motive, in which the purpose is to efficiently produce a very precise product, provider, or cease end result. This entails ensuring that the scope, time, coins, and assets are getting used inside the confines allowed. The task supervisor accomplishes this purpose through making plans, organizing, motivating, and controlling belongings to gain dreams.

The topics that I will go through are what I've placed to be the crucial thing elements of a success undertaking manipulate. There is not a completely unique order to those and now not one is necessarily extra important than each unique. Project manipulate encompasses all of those dispositions at numerous times inside the route of a project on the way to obtain fulfillment.

Chapter 7: Be Responsive

As a mission supervisor you may must be responsive in a timely way. This is one of the most essential factors of mission manage. When you or your organization is shriveled to address a mission, the patron and/or your organisation desires to recognize which you're actively available to cope with issues and solution any questions or issues that they'll have. This is first-rate than efficiently speaking, in case you want to be illustrated in a later phase.

Being responsive does now not imply that you have every option to every query right away. What it way is that you are returning mobile phone calls and emails, or while a consumer involves you with a query or trouble which you're capable of get once more to them inner a properly timed manner. Even in case you are responding to them and permitting them to recognise that you'll want to get extra facts and get again to them. Just via responding to the customer and allowing them to realise which you are available and inclined to assist them and solution questions can relieve feasible tensions inside the courting. This is what your client or boss will respect the maximum.

When a purchaser calls you want to take your customers name even if you do now not have any solutions for them. It is brilliant to let them vent and no longer interrupt them till they will be completed, typically they just want to sense they are being heard. Try to concentrate and recognize their feelings and goals. I should advise having a pen and paper available so that you can take notes. When they have got finished venting, reply to them by using the usage of the use of repeating the concerns and display empathy. If you have got got answers allow them to recognize as you respond, however if not tell them that you are both operating on a solution or it's miles a few factor that you may take a look at.

Ex 1. Your boss has really came about a latest mission to you. He informs you that the purchaser truely known as and wants to apprehend at what degree we're inside the undertaking. Of route you determined, I have become genuinely given this assignment and I don't comprehend a few detail about it. Should you without delay name the client and say which you have been just given the project and you don't recognise what goes on but. The solution ought to be no. The higher approach might be to spend 5 or 10 mins reviewing the records and getting a simple idea and then giving them a name. The reaction need to be: "Hello, my name is John/Jane Doe and I may be the venture manager for this assignment. I desired to touch base with you and can help you recognise we're within the method of planning your mission. I will keep you informed and up to date to a tee; I apprehend you may be impressed with the outcome." This easy cellular phone call will usually offer you with about 24 hours of breathing room to without a doubt get a draw near of the state of affairs. After this smartphone name you can commonly supply updates with the useful resource of e-mail as desired.

Ex 2. The venture has professional a few sudden problems, which through the manner happens more instances than task managers are inclined to admit. The patron is asking you to discover the reputation however lamentably your group has dropped the ball and you haven't any statistics to ease their troubles. Many humans would pick out out to dismiss the calls till they've got some news. This is clearly the wrong manner to deal with this. You need to preserve the purchaser up to date simply so they recognize which you aren't preserving off them. You have to either telephone or electronic mail the client with a first rate replace. You need to sincerely kingdom what has been successfully completed. Then you may allow them to understand the subsequent few steps. While reassuring them that you may update them even as your organization has a plan of attack.

Chapter 8: Micromanaging Efficiently

In any undertaking you can have numerous agencies of humans that you need to paintings with in order to complete the duties. You will have to consciousness on many regions. Once you've got a plan and feature decided who you may be delegating art work to it is great to provide them suggestions and a time frame. Let them understand that the door is open for any questions they may have. Have a assembly to allow your group be involved on how they revel in they could awesome accomplish the responsibilities. You need your group to prevail so that you need to provide them the freedom to give you new and stepped forward strategies to perform responsibilities, but you need to in the long run be the choice maker on how those responsibilities are executed.

During a challenge you need to have checkpoints. You need to recognize if a venture goes outdoor of the scope, fee, or time frame. A well assignment supervisor desires to permit personnel to excel however despite the reality that be privy to what's taking location spherical him. A proper supervisor may additionally even want to test in and make sure that the pointers supplied within the starting of the task are although being accompanied in the time-frame required. And even as you discover that a person to your organization is strolling outside the scope or getting off beam you need to manual them lower back in the perfect way. Your task is first-class as robust as your weakest hyperlink.

Ex 1. You are in the middle of a big task and you have were given positioned that a member of your organization has decided to take quick cuts due to the truth he is falling not on time. The appropriate response might be to have a communicate approximately the scope of the mission and wherein they're. You will need to decide at what point they began breaking manner which altered the mission plan. The nice method then is to be wonderful and permit them to understand that they are valued. The next step would be to provide an reason of to the worker once more what step they unnoticed in the assignment plan and why it's far this form of crucial step. And then I should ask them how they advise we're able to remedy the state of affairs. It is commonly terrific to offer an employee a chance to figure out the way to restoration their non-public mistake, time allowing. It is a boom possibility.

Chapter 9: Keep Evolving and Improving

Kaizen is the Japanese time period for improvement. It does no longer rely what enterprise you're in you need to maintain to hold up with the current inclinations. That approach that you need to recognize what the modern traits to your employer are. This is some aspect that you need to placed effort and time into on a each day foundation. There will continuously be a new and progressed product or approach as a way to make all people's lifestyles an awful lot much less complex. Working on improving and developing your abilities for only a few mins a day ought to make the distinction among and not unusual supervisor and the outstanding manager.

There are many resources that you could use to maintain present day on your company. The net is the brilliant place to begin. It can be very clean to use the internet to enroll in blogs and be part of up for webinars. Then, of route you can use extra traditional strategies together with reading the modern day books and agency guides.

There is a pronouncing in IT and this is that you are constantly seeking to hold up with the Red Queen. This approach that whilst you in the end seize up and have a look at the whole lot you observed you need to understand it's miles already preceding information and you will want to research some factor new. Thankfully maximum fields aren't as evolving as generation, but you continue to need to hold to have a look at and evolve or you may be left at the back of for your industry.

Ex 1. Let's say which you are a Project Manager for a advent corporation. You want to be inquisitive about gaining knowledge of as a extremely good deal as you could approximately lean manufacturing, layout manipulate, safety regulations, and of path new product presenting with a purpose to boom income and hopefully growth productiveness, whilst reducing fees, and time. The first thing you may do is go browsing and are searching for boards and books. There also are social networking sites that have groups that similar people on your profession are constantly updating records and developments. Knowing at the same time as a cutting-edge code has been mandated will prevent costly errors and enhance your reputation.

Chapter 10: Detailed Organized Planning

A essential element of venture control is fine prepared making plans. You should be very unique for you to effectively follow via with a project. Proper making plans of a challenge includes organizing the records. The first step is to recognise what you are trying to carry out, the tool you can want, and the time-frame you're jogging with. Then you need to create a agenda so that you have a place to begin, the steps that you want to increase so as, and an completing purpose or what you would really like to appearance because the very last final results. It is a good idea to think about viable snags to your plan and permit for added time for those unexpected activities.

One of the excellent tool for assistance with that is the Gantt chart. The Gantt chart basically lists out all of the steps that want to be accompanied via in a task. Also on each line you can put in an estimation of time for every hobby and list whether or no longer this step is vital. Some steps you need to correctly entire with the intention to flow into on. A software program software program that is very popular to help you listing out each step is Microsoft Project.

Ex 1. In this case we are capable of take a small assignment and come up with a task plan and time body. The assignment can be to install a trendy door knob and the time frame can be 20 minutes. Each step need to be finished correctly in advance than you can circulate onto the subsequent.

The steps are as follows:

1.   Prepare for installation. Est. Time 5 minutes.

2.   Have replacement door knob.

three.   Have gear: Screwdriver.

four.   Unscrew previous door knob. Est. Time 3 mins.

five.   Remove hardware from door. Est. Time 2 minutes.

6.   Remove hardware from packaging. Est. Time 1 minute.

7.   Install new hardware. Est. Time five minutes.

8.   Test new hardware to ensure that it nicely works. Est. Time 2 mins.

nine.   Clean up Est. Time 2 mins.

Chapter 11: Work With a Sense of Urgency

This step isn't some component this is truly mentioned lots but is critical. The enjoy of urgency is definitely clearly an thoughts-set you are taking on all through a project. This is largely having an thoughts-set that what you are doing is crucial and that right away motion is needed to finish your responsibilities. This is essential because so you can meet deadlines and stay heading in the right direction you need to have a clean vision of in that you are going and be determined to live focused and make it through each step in a timely way.

A sense of urgency isn't demanding and does no longer require you to feel anything poor. This must be a immoderate high-quality experience in which you are concentrating to your desires and running constructively in a nicely timed way to move from one venture to the subsequent. You should have the mind-set that your challenge is pressing, well communicate, and use your sources.

Note: Not each venture that you are attempting to complete may be urgent and if you try to deal with them as they're, then you will kill productiveness. This is in which the exceptional project manager will stand out because they'll have the ability to distinguish the among an urgent challenge and a non-urgent task.

Ex 1. You have truly been assigned a current mission. In this undertaking you need to affect your boss. Don't be anxious and don't be harassed. The very first element you want to do is to offer you your project plan. This should be a little by little guide on what you are attempting to reap, the tools used, a timetable, and what you need can be the prevent very last results. The largest a part of strolling with a experience of urgency is that you preserve on top of troubles and respond in a nicely timed manner. Let's say that your boss is seeking out an replace and you have vendors which can be procrastinating. You want to permit the groups understand that you are on a agenda on the same time as accumulating as thousands records so you are aware about where they'll be within the method. Then you may want to arrange this records proper away so you are capable of deliver your boss or patron an update. Even when you have bad data you still need to maintain them up to date and allow them to understand what you advocate on doing to maintain transferring ahead. This is the way you figure with a feel of urgency.

Chapter 12: Visualize the Outcome You Want

In every assignment you can normally start off with an give up goal in thoughts. In order to acquire this aim you're going to must visualize the manner you see the venture gambling out. This is a very important step in planning. You can not absolutely say that you need to get from A to Z but the data do not bear in mind. The statistics are realistically how you offer you along with your project plan.

There are a number of human beings talking approximately visualizing your future and the way you're ultimately the simplest if you want to make all your dreams come to fruition. This is particularly true because what you be aware is what you get. The identical is going for a assignment. If you be aware a assignment and do not surely be privy to the information you could have a totally narrow challenge plan and opportunities are that it'll no longer pop out as some thing aside from the smooth way you noticed it. On the opposite hand, if you test a undertaking and visualize what you would like the challenge with a view to accomplish for you, it is going to be greater in-intensity and you're planning with be better organized.

Ex 1. Your commercial enterprise organisation wants to set up a new server device because of the fact your vintage server warranty is ready to run out and also you observe your modern-day-day server has been having greater problems than normal in recent times. So you start off thru the use of visualizing what you would really like your server to do. In this case you would love the server to be in rate of walking your net internet page. So you visualize the internet site which you would like to have and the talents you would really like it to finish. For instance you are a nonprofit enterprise so you want your internet web page to be a view into your company. You furthermore determine which you would love so as to collect donations for your net website because of the fact donations are how nonprofits stay on, meet financial dreams, and duties, at the same time as however offering for its meant use of showcasing your business employer enterprise. Once you have got your imaginative and prescient you will be capable of contact the distinct groups and nicely communicate that vision as a way to make your imaginative and prescient a a success fact.

Ex 2. On the alternative hand permit's say which you're a nonprofit that would like to start a internet website online on your corporation. If you don't have any imaginative and prescient it will be very mounted and vain. Possibly even a web web page with some pix and a few terms about the company. You should don't forget that you will best get what you observe, so in case you see a fundamental net website online that is all you can get. If you observe a net website online that has a few meat to it and allows you to capture a vacationer's hobby and get keep of donations, you could have that. That can be very viable with proper planning. This is in which you could see that there can be without a doubt an advantage to visualizing what your very last product might be.

Chapter 13: Communication Is Key

Communication isn't always nice important for initiatives however it is also vital to any form of achievement in lifestyles. Everybody will now not be on the identical net web page however with conversation they'll be strolling to go lower back to the identical surely stated endpoint of a project. Everybody may additionally moreover have excellent mind on how they may acquire that endpoint and that is why conversation is so vital. You will want to be smooth on the manner you assume to come back to that endpoint and what steps have to be taken to get there.

Good communicators speak truely. Many put together what they want to say earlier of time and anticipate earlier than they talk. They explicit in truth what message they need to deliver. They ought to verify that the opportunity person is privy to. This generally completed via having the person repeat another time to them how they interpret the information. Identifying and emphasizing key factors is important. The cause need to be well described. Don't end up emotional, feelings sincerely cloud your thinking and the way you specific your self. It is also useful to install writing stuff down so you have to have a reference at a later date. Keeping nicely notes will constantly be a advantage for a a achievement assignment very last outcomes.

Ex 1. Your modern-day mission would require a mission assembly in conjunction with your organization. Before the assembly you could need to plan out exactly what you want to carry out for your meeting. You begin the assembly writing down what your undertaking is. Then you determine in conjunction with your organization to offer you with a plan at the way you get out of your start point all of the way in your stop thing. It may be very beneficial if you write down your desires and steps in an organized way. I pick out to use a white board so that everyone able to view and visualize the equal information. With this technique you can communicate all devices along side your organization and ensure that everyone is aware about the tool and the last very last effects. When your meeting is entire everyone should be clean on what the desires of the mission are and what their roles are all through the challenge.

Ex 2. You have a large assignment and you decide to deliver out an e mail to all people listing the task desires and what each person's characteristic is. This is the worst way to talk a large task. First, of all maximum of the employees will now not look at the email or recognize the content. Second, your plan will not have input from certainly one of a type key individuals. If it is not viable to get each person in a room together then no much less than you want to preserve a conference call. It is pleasant to deliver out an e-mail earlier than the choice with a spine of a project plan. Once the decision has ended another email must be despatched that summarizes what became noted and determined on at some level inside the call. This summary need to encompass an up to date challenge plan that was hopefully finished during this convention name. This way every body is privy to what to anticipate.

Chapter 14: One Step at a Time

Have you ever heard the section, "don't placed the cart within the the front of the horse"? The that means is simple, while you're taking on a venture you do now not just snap your fingers and it's miles completed. This is a very essential idea. Once you are organized to start your undertaking you may need on the way to paintings on your mission in steps. It is often a continuing way which you frequently accomplish. Most of those steps will want to be in a splendid order based totally totally on the project desires.

There can be issues that rise up in the course of a venture and you can even want to move once more to preceding steps to repair an trouble or add extra steps. The Gantt chart is incredible tool for the grade by grade plans in assignment manage. A Gantt chart is a bar chart that lets you visualize your venture time table with the aid of supplying you with a clean instance of the mission repute.

A Gantt chart uses a horizontal axis to symbolize the whole mission time span. It then breaks the stairs down into increments of days, weeks, or months. The vertical axis represents the duties that make up a task. Gantt charts can be smooth with just a few steps or comprise loads of steps. The pleasant problem with Gantt charts is that they may be made for use for pushing forward and not very forgiving transferring backwards or unilaterally.

Another tool that I truely have decided to be useful are drift charts. A flow chart is a graphical or symbolic illustration of a machine. These charts are very useful for smaller responsibilities or breaking apart massive duties into smaller sections. Flow Charts are specifically beneficial if you have steps that could bring about different possible effects.

Ex 1. Your boss has just asked you to create a number one task plan for present day day hires that can be done at some level inside the organization. This is most effective a clean chart that may be created in Excel. The steps listed are simple but there is a wonderful order that want to be determined. For Example, there is no element in starting the primary steps when you have not hired the worker. Also you need to entire business enterprise and tax files as a way to set them up on your payroll device due to governmental necessities. You can skip round to some extent at the same time as growing a folder and its contents. Computers also permit for versatility of being able to skip spherical among responsibilities any time after a modern-day employee has been employed. Many of these steps are not consistent as to what order they have to be in.

Chapter 15: Risk Management

Risk manipulate is one of the toughest duties that you may be wanting to achieve achievement as a undertaking manager. It may be exquisite if every venture had a extremely good assignment plan and every step flowed resultseasily into the others without any issues. The fact is that a assignment plan is like your street map and greater often than no longer you can face problems that require adjustments for your plan. My goal isn't to get into the professional necessities of risk control but to truely provide you with a primary knowledge.

When you are managing a project your method is to have a first rate project plan and examine the first rate steps to try to understand any possible threats. You then need to determine the hazard of those threats and reduce and manage those dangers as masses as possible. In a few times you've got risks which you aren't able to decide after which at that element you definitely want to art work collectively along side your team to decide out answers because the problems rise up. When being reactive in threat manage you can really want to recognise what opportunity expenses are associated with every solution. Although it can be over budget the most inexpensive answer is not continually the first-rate and in plenty of times it will simply price greater money and time in the long run if you try and lessen corners. So make certain you evaluation all your alternatives and select out the handiest on the way to be the maximum critical benefit to the overall task.

Ex 1. You are riding in Chicago from the suburbs permit's say to North Ave. Beach. You have your GPS set up to take you on the maximum direct path. When you begin riding you hit pot holes. No huge deal you sincerely maneuver around them. This is an instance of a minor trade you could want to make if you need to get in your vacation spot.

EX 2. When you are halfway for your vacation spot you hit the regular Chicago manufacturing area detour. That has simply thrown a wrench into your plan. In this situation you will observe the detour signs and symptoms and symptoms and symptoms and get all over again in keeping with your plan. Once you gain the seashore the very last outcomes may be just like if you took your authentic route however the possibility rate of getting to the beach is what has modified.

EX 3. You may additionally moreover have hit a pothole and ended up with a flat tire or you may emerge as using extra money in fuel then you definately truely surely had budgeted. This is a clean analogy however the premise is the same. You perceive the threat and make adjustments as wanted considering the possibility cost. Time Management

So many factors of a challenge are crucial, however time manipulate is usually the most important element. While a mission is inside the making plans tiers there may be a choice at the assignment prevent date. That way that each one of the worried activities have agreed on a date that this undertaking need to be finished through manner of. The one hassle which you ought to understand is that during case you are not capable of manipulate some time nearly on every occasion you burst off your time table you can cross over price range. These too additives work hand in hand.

When a assignment plan is created you may have dates for every step along the manner. It is crucial to hold to the time desk. There will usually be constraints that have an effect on time control, the restrictions are the sources. These can be resources which incorporates: inventory, difficult paintings, and out of doors professionals. You will want to artwork with a enjoy of urgency at the manner to coordinate all the sources a high-quality manner to maintain for your time desk. Finish a task early a purchaser will in no manner complain, but in case you run behind the schedule that you laid out for them they may come to be very annoyed and query your competency. It is usually better to promise much less and offer greater. No one has ever complained that their expectancies were handed.

In the begin of the challenge you can have with a chunk of achievement achieved your due diligence on your risk control. During that thing you may generally provide you with a buffer of time for any unexpected problems. I in my view like to apply a 25% buffer. This buffer is your coverage if you have any problems, it'll provide you extra time to paintings on them.

Example 1. You have a challenge assembly scheduled for inner non-public and out of doors carriers. The most vital seller to the task has to hold postponing the meeting because of agenda conflicts. In this example whenever you want to reschedule you're running out of doors of your high-quality time body, except you have got constructed in extra time. This will boom the expenses of the task. If you find this is taking vicinity you'll need to achieve this to get this seller on board and make you a subject.

Example 2. There is an internal meeting and some new facts has come to moderate. You will need to regulate your venture plan to encompass this records. The enterprise estimates that there may be a further 18 hours of exertions with a view to want to be included but they permit you to recognize that you will need to finish the project inside the true time body. Of direction, in case you at the beginning quoted the customer one hundred hours you should already have a 25 hour buffer constructed in.

Your reaction to the purchaser have to be "we are capable of see what we can do and are to be had as close to as viable". You will then want to allow your group understand of any change due to the fact it could change the scope of the assignment and there may be a similarly fee.

Chapter 16: Stay In Budget

Any assignment you adopt will have a finances. The fee range is an in depth estimate of all prices that undertaking the challenge will incur. The not unusual finances includes, however is not constrained to: exertions, materials, taking walks fees, and different direct prices. The finances enables to control expectations with the resource of the use of the remarkable estimate. As we stated with the time control if you have a exchange of scope it will in many instances have an impact in your price range and further charges also can want to be assessed.

It may be very easy to move over fee variety on a project and while you start going over budget it's going to generally have a snowball effect. The quickest and yet the sneakiest manner to move over price variety is to postpone or make bigger the timeline of the task. If you enlarge the project due for any purpose you can incur greater undertaking control hours. This is in which once more hazard control comes into play. When you first be conscious that a venture goes over finances and it'll change the scope the great issue to do is to Stop, Drop, and Roll. Stop your undertaking, drop, not actually but back down the project, and roll, proper now get along with your institution and control to assess the scenario and make choices on whether or not to move ahead or begin a verbal exchange collectively together with your customer.

Example 1. You are working on a project and your consumer realizes that they ought to placed the project on preserve for some weeks. In this few weeks you'll want to test in with the purchaser so one can preserve an open line of conversation so that you are able to resume the undertaking. This verbal exchange will upload greater hours on the way to jeopardize the fee variety of the assignment. Usually the hours will not be significant however it's going to probable be a few aspect to speak approximately collectively along with your institution.

Example 2. You are putting in a contemporary electrical area and you examine that some wires aren't as a lot as code. This come to be now not part of the estimate and the wiring isn't part of the original quote. This is the time you Stop, Drop, and Roll. You should immediately forestall the task and meet with the proper activities as a way to assess the available options. In this example there genuinely are not any options, so you need to be prepared to get the purchaser a quote on the today's wiring just so the distance is as much as code and they're able to skip any required inspections.

Chapter 17: Deliver a Quality Product and Service

The key to a a achievement agency is to supply a first-rate product and provider. If your product and enterprise is in reality without a doubt worth the cash clients will pay. If your services or products are sub-standard the customers will pay extra for your competition to get the excellent they need. Quality is what it all comes down to. All round extraordinary must be at a immoderate diploma on the way to maintain customers and increase your consumer base.

When you buy a company commonly the charges are typically around the same range among opposition. The purpose that you choose out one provider over every other is that you may collect a few element that you are able to receive. Everyone has minimal requirements that they may be inclined to absolutely accept if the charge is right. As prolonged as the ones requirements are met you could keep coming again for greater. When you need a higher excellent you're inclined to pay greater. Most groups realize a way to keep the quantity slightly above those minimal necessities to keep developing and hold clients coming returned for additonal.

Example 1. There are many groups of cell phones. Many companies use the same cell telephones so the great is precisely the equal. What differentiates the mobile cell smartphone corporations is the provider. Some people are satisfied to pay the lowest fee viable to have a few spotty carrier. While others need a higher provider in order that they pay a organisation even greater. If you have got dropped calls and tour to regions that don't get company you may most probable be interested in paying greater to get a higher provider.

Chapter 18: Multi-Tasking

Multitasking is the capability if you want to switch halfway from one project to some other and then again again. Many human beings recollect multi-tasking to be the capability to cope with a couple of issue at a time, in reality this isn't feasible. Looking at pc structures for instance, most laptop systems are also unable to address multitasking except they have got a multicore processor. What is viable is that you could skip to and fro among initiatives and obligations pretty fast which permits you to revel in as in spite of the reality that you're correctly multitasking.

In task control you may be working on a challenge for one purchaser and you may get a call from some different customer. You could have a very good manner to save you what you are doing to interchange to each distinctive hassle. Then you may want to decide which trouble ought to take priority. Once you have got got determined on the client priorities and what wants to be finished for whom, you can typically be going backward and forward at the manner to cope with all issues, for every consumer, as they rank on your priorities.

Example 1. In the center of a big challenge with a last date this is drawing near fast, every different purchaser calls with a want for urgency and now you've got got were given obligations that need to be finished with the aid of Friday. In order to accomplish each and hold up collectively along with your cause dates you could have milestones that you may be looking to reap. This generally consists of finishing a undertaking for one purchaser and then moving to a project for a few different client. Sometimes you may want to save you halfway thru a project as a way to accomplish a task for a few different customer due to urgency of that challenge.

Chapter 19: Understanding Project Management

What is a Project

We're going to discover about essential necessities. First, we're going to investigate is what exactly is a undertaking and secondly, we're going to discover approximately what's a deliverable.

So first, permit's recognize what precisely is a venture. A assignment is a temporary endeavour. This way that it isn't always some thing that is ongoing, it's far a one-off try.

Secondly, a venture is a few aspect that has a real beginning and an stop. So, it has an area to begin and it has an completing thing.

Lastly, a venture is something which creates a completely precise product company or stop result. The emphasis proper proper here is on the problem of a project being precise and handing over something this is genuinely one in every of a type.

- A building

- Software software (e.G. IPhone or Android app)

- A Website

- A bridge

How do the examples above encompass the dispositions of a task?

Let's take the example of a bridge in a metropolis. Firstly, constructing a bridge is a one-off project, you can sincerely should build it as fast as (with a bit of success), and it's now not some component you'll should artwork on each day, so growing the bridge is a temporary occasion. Also, it is specific for the reason that bridge has a particular starting and finishing element because of the reality it's far connecting precise factors of the town. Furthermore, it has a completely unique format to make it aesthetically attractive to examine. The equal standards can be without problems applied to the relaxation of the examples I've stated, recall how and why they can be taken into consideration projects.

Lastly, I'd advocate you exercising the equal requirements to the art work you do, what type of tasks have you worked on?

Next, allow's take the instance of a building as a challenge. We will use this to provide an reason behind a few different crucial issue of mission control, known as a deliverable.

What is a Deliverable

The specific very last results it's produced via the challenge is known as the task deliverable. This can encompass any products you may boom, or offerings, or any shape of prevent end result that came about because of doing the project. Therefore, a constructing in Manhattan, New York is a mission. This is due to the fact every constructing can be built in a very specific place that no outstanding building can occupy, it will have its very very own specifications and requirements for raw substances and labour.

A Deliverable also can be taken into consideration due to the fact the Output of the assignment, i.E. What comes out of the venture as quickly as you have carried it out.

It is also critical to apprehend that a challenge furthermore has deliverables inner itself. For instance, the deliverables inside the building task may be the flooring, pillars, and the roof of the shape you're developing. The very last deliverable of this undertaking might be the overall completed constructing. On the opposite hand, the entirety that went into developing that constructing were moreover deliverables which have been internal that task.

Therefore, we are able to summarize the building venture as follows:

Final Deliverable: A Completed Building

Deliverables Within the Project: Floor, Pillars, Roof

Chapter 20: The Project Manager's Role on a Project

Who is the Project Manager and what does he/she do at the project?

Project Manager = PM

The mission manager's position is a large one. It can not be simplified into a person particular project. Instead, think about the undertaking manager because the hold close orchestrator of all that desires to be completed to complete the undertaking. To simplify topics absolutely, the number one venture of the undertaking supervisor, is to speak. The PM is the center round whom everything else takes region and is the most effective character who's aware about what is taking region on precise elements of the assignment.

Let us convey clarity to the precise position and responsibilities of project managers and the movements they may be concerned in. We will examine what's predicted parents and what we should be doing on the same time as dealing with tasks:

Success or Failure

A Project Manager is the individual that is responsible for the challenge, without a doubt positioned, you're the undertaking champion. You are the person selected via your corporation/industrial agency employer to fulfill the assignment's desires and to govern the mission team. The PM is responsible for the success or failure of the challenge, this make or harm responsibility rests on the PM's shoulders and is the bottom line in their commonplace performance.

Communication Skills

It might be very important for the PM to have first-rate communique abilities. Over 90 percent of a PM's time is spent speaking among human beings, agencies, and stakeholders. Your coordination with stakeholders is primarily based for your functionality to apprehend, extend, and percent relevant records at the right time to make sure the project is going at the proper song.

Leadership, Training, Negotiating, Coaching

You can be predicted to spend a while appearing manage through the usage of providing path to the assignment. You will construct organizations, get them to art work together successfully and provide the vital schooling to growth their effectiveness. Also, you will be indulging in negotiations to acquire the great sources for the challenge. In case the undertaking runs into troubles, you'll be coping with war to remedy problems. Your institution will depend on you for education and guiding them through the venture.

Coordinating and getting paintings performed via others

A sturdy mission supervisor is the last coordinator of human beings and property. This is because of the truth you're responsible for getting paintings achieved through unique human beings for the challenge. You coordinate the artwork amongst unique departments, stakeholders and undertaking institution. I favored to reiterate and highlight a key component proper right here, bear in mind that the project supervisor isn't always liable for doing the challenge work with the aid of the use of the use of himself/herself. Instead, you get the mission artwork done through exceptional human beings and manipulate that work to make certain it's miles carried out the first-rate manner and fulfills the requirements.

Developing time table, charge variety and reserves

Given that responsibilities function in strict timelines, the PM is responsible for growing a realistic agenda. We say practical seeing that we want to allocate an low priced time for the manner extended it's going to take to perform each interest concerned in doing the project paintings. You can't be too stringent with the timing and neither are you able to be too generous with it, it desires to be actually proper, which means that sensible.

Next, growing the challenge budget is finished through manner of summing up all of the expenses on all of the sorts of paintings concerned at the challenge, and this is your challenge fee range. In other terms, add all of the expenses of all of the form of art work required to do the venture, and you could have your finances. (Explained in greater element in Cost Management Chapter)

Furthermore, word that tasks can go through adjustments or disturbances that may negatively have an impact to your fees or time desk. You can't have sufficient cash to transport over your undertaking price variety and neither are you able to have enough money to have delays. To cope with this, mission managers are chargeable for growing reserves.

Reserves are like protection backups of time and money. The assignment manager allocates bits of more money and time on the undertaking to keep away from deal with disturbing situations or troubles confronted via way of way of the challenge. So, if your challenge is getting no longer on time with the useful resource of 10 days and if you have already got a time reserve of 15 days, you may despite the fact that manage to pay for to go through that put off as it will although meet the project cut-off date date.

Note that the reserves of time and money are blanketed as a part of your finances and time table and are not some thing you add in some time even as a problem takes location at the task.

Managing Human and Material Resources

Projects hire skilled professionals and specialized tool. The task supervisor is liable for figuring out what resources are required for a assignment. On your non-public, you will be allocated remarkable assets inside the form of a project team, however, tasks can be complicated and require multiple skillsets and belongings. It is likely you'll have to build up property from unique components of commercial enterprise employer and from outdoor companies as the task advances.

Your motive is probably to get the most awesome belongings to paintings in your project. As a cease end result, you may find out your self negotiating with department managers e.G. IT, Marketing, Engineering and lots of others. To borrow their group individuals.

Furthermore, you may create procedure descriptions for project organization individuals and stakeholders at the side of making sure that roles and obligations are clearly assigned.

As the project paintings receives performed, you may music the overall overall overall performance of the group individuals and look at the want for training. You can also create rewards and popularity structures for engaging in venture milestones to keep the morale and widespread universal overall performance ranges immoderate throughout organization contributors.

Performance Tracking

Since it's far vital to tune the improvement of the venture, you are anticipated to degree task ordinary common overall performance and become aware about any variances the usage of specific commonplace performance metrics. Hence, earlier than you start off collectively with your venture, it's far crucial to determine thru what technique or measures will you use to assess the general overall performance. E.G. Is your undertaking on growing a present day vehicle engine turning in the excessive-tempo miles/hour or Km/Hour values that the engine grow to be to begin with bear in mind to deliver? So earlier than you begin the challenge, define the overall overall performance metrics.

Responding to Changes – Analyze Impact and Recommend Corrective Action

In nowadays's rapid paced surroundings of fast product launches and ever evolving marketplace conditions, the challenge supervisor is expected to respond brief to changes. Changes may be huge in nature, e.G. Client has requested for introduced product talents, the challenge cut-off date has been decreased, the mission fee range has been lessen to seventy five% of the unique amount and so forth.

It's critical to be aware that changes are not carried out right away without evaluation at the mission. They are analyzed, after that are either commonplace, or rejected. (Discussed in greater detail later)

The mission manager offers with modifications on the undertaking with the aid of reading the effect of the adjustments (what will be affected because of the trade and through how an awful lot) after which recommending corrective moves (What can we do to set the venture on the proper tune in reality so it however achieves its aim.

Chapter 21: How tasks are Started – What is the Business case

A enterprise business enterprise case offers the reasoning for beginning the venture. It explains the cause why a venture became initiated within the first region and why it is properly worth taking on. It offers information on whether or not or not or now not it is profitable making an investment on this undertaking, whether it makes enjoy to spend money on this assignment and will it supply the organisation the first-class returns.

Projects are at the whole taken up to satisfy the goals of corporations, that could range from catering to marketplace dreams, to technological enhancements. For instance, an car manufacturer may additionally moreover additionally need to release a ultra-modern product line of motors to boom profits and as a quit result income and profitability. Or it is able to need to growth its productiveness by means of way of upgrading its production unit system, so it could produce extra vehicles in lesser time. All those motives can be taken into consideration because of the truth the employer case for beginning a assignment.

.

What are Process Groups

The manage of any form of a assignment may be divided into five awesome technique organizations. Think of it due to the fact the drift or series of stages that your mission wants to go through. Whenever you start a task, try to method it as in case you are going to section its improvement through specific levels.

How does the use of method agencies assist us?

They help us set up the task and make it an awful lot much less difficult to govern. Things begin to make masses extra experience even as you placed everything in its proper location. Large duties may want to have severa variables, and if you don't put them so as, you would possibly threat challenge failure. Without order, responsibilities can go into chaos.

- The method businesses encompass:

- Initiating

- Planning

- Executing

- Monitoring and Controlling

- Closing

The challenge manager uses the ones technique companies which will control and supervise the mission.

Now allow's try and understand every of these exceptional technique groups one after the other.

Initiating

Initiating is especially concerned with beginning a challenge and growing the preliminary requirements as a way to start the task. So, starting is about how you're going to start off with the mission inside the first place and what precisely are we able to require to make the assignment a fulfillment. Think of it as a completely wide degree evaluate of the entire project.

Planning

Planning is in that you want to installation what are the things you want to do as a manner to get the venture performed. This is in which you may outline the route of movement on how the mission's desires and necessities are going to be done. It is typically encouraged to spend extra time in planning, the more idea you located into the planning, the smoother can be the execution and the lesser troubles you may come across. Think approximately the manner you're going to get the work completed on the mission and what risks can the challenge stumble upon. If you've protected the ones factors in the early levels, then you gained't need to face problems later due to loss of making plans or education.

Executing

Executing is worried with getting the real paintings finished at the mission as a manner to achieve its goals. This is in that you're getting paintings completed thru human beings and thru material sources.

Monitoring and Controlling

Monitoring and Controlling consists of monitoring, reviewing and regulating the general performance of the challenge. It is involved with ensuring that all of the paintings that is being done on the project is on the proper tune. In case there are any problems which is probably taking place at the task you need to make certain which you're imposing corrective movements that allows you to bring the challenge lower again heading in the proper direction as quickly as possible.

Monitoring

When it entails tracking, this in particular refers to gathering, measuring and then analysing facts related to the mission's overall performance so that it will understand versions. Variations check with something that isn't always getting in accordance to devise at the assignment.

Controlling

Controlling involves ensuring which you're taking the right and appropriate corrective actions on the way to hold the assignment fashionable usual performance lower back heading within the proper direction. In case you be aware that the general normal performance of the project goes off target, you want to put into effect the right corrective measures and make the right modifications as consistent with the original plan in an effort to make certain that the artwork is being finished steady with requirements and

Closing

Closing is all approximately finalizing all the sports activities sports and the artwork this is being finished on the assignment and formally remaining it. This is wherein you can wrap up all the artwork finished on the task.

Process groups assist deliver order in your venture. Think of these way agencies as a template which help you set up a mission via its one-of-a-type tiers.

What are Knowledge Areas

This is a few exceptional one of the most crucial topics as regards to undertaking manipulate. In smooth phrases, understanding regions are like commands of know-how primarily based mostly on which tasks are managed.

For example, every venture faces dangers, each task has a term/time table, expenses, stakeholders and so on. Hence, as a manner to help manage those notable elements of a assignment, those elements are categorized as facts regions.

Each know-how place has its very own strategies. The techniques inside each information place embody the pointers via which a particular detail of the mission is controlled. For instance, the Risk Management information place consists of all of the techniques and tips which show the manner to manipulate Risk on a project. The identical applies for the rest, e.G. The price manage information region includes the processes and pointers which show you a way to control fees on a mission.

Overall, Projects are managed sooner or later of 10 precise expertise areas which encompass:

- Integration Management

- Scope Management

- Schedule Management

- Cost Management

- Quality Management

- Resource Management

- Communications Management

- Risk Management

- Procurement Management

- and Stakeholder Management

Let's recognize what each of these regions imply (Please be conscious we will virtually cover all of those records areas in more element later, for now, it's miles very critical to get an overhead perspective on what function does every play in phrases of dealing with initiatives.)

Integration control focuses on Integrating Processes for the duration of the project

It is involved with identifying what shape of exertions is needed to do the venture after which integrating strategies at some stage in the mission with every exceptional. Think of integration control as a manner to make the best of a type additives of the project paintings together.

Scope Management specializes in Developing Product Scope and Project Scope

For product scope, it covers what is blanketed within the product you need to increase and what now not to embody in the product.

Similarly, for assignment scope, it covers what have to be covered inside the artwork of the assignment and what is not blanketed in the assignment artwork.

Schedule Management makes a speciality of Developing Project Schedule

It includes growing with the timelines and the Project Schedule. It moreover includes identifying how a good buy time each interest at the undertaking will take.

How an lousy lot time is that this mission going to take? When do we start it? When can we give up it? How an awful lot time is it going to soak up order to finish the tremendous sports which may be involved inside the mission. So you can use all of that statistics to increase the agenda of the undertaking.

Cost Management is all about estimating the costs that allows you to be incurred at the task. It specializes in Estimating fees for human assets and cloth assets

For example, when you have engineers running on the undertaking, then what are going to be the costs on the mission nearly approximately human property or the people running at the challenge.

It moreover consists of developing with the price of fabric resources, which includes the price of the material used at the mission. E.G. If you are using one-of-a-kind types of materials as an example in case you are constructing a constructing, what is going to be the price of the metal and cement.

Quality Management makes a speciality of Ensuring Product Quality and Process Quality

Firstly, for product awesome, it ensures that the pleasant of the goods being created is at the best degree.

Secondly for method exceptional, it guarantees that the strategies of task manipulate itself are at the suitable stage.

For instance, if you're growing a software program program product, then you definitely simply want to make sure that the satisfactory of the deliverable, which means the product outstanding is of a sure wellknown. You additionally want to make sure that the techniques getting used to make bigger that IT software application program product are of the proper incredible.

Therefore, each the great of your techniques and of the products that need to be of a excessive pleasant fashionable.

Resource Management focuses on Acquiring, Developing and Managing People and Material Resources

It consists of getting the proper human beings to paintings at the task, growing them, after which coping with their common overall performance.

It also entails getting the proper material assets and device to do the task art work on the right time whilst it's miles required at the project.

Communication Management focuses on Providing Right facts to the right people at the proper time

It guarantees that the right form of facts reaches the right kind of people in a well timed and green manner.

Risk Management makes a speciality of Dealing with Opportunities and Threats

Risks can be of different types. You could have high-quality dangers which might be known as opportunities and then you can have horrible dangers which can be known as threats.

You want to make sure that opportunities on the undertaking are capitalized on and threats to the project are taken care of. This includes growing particular plans to address possibilities and threats.

Procurement Management specializes in Purchasing Goods, Managing Contracts

It consists of buying items or offerings which you require on the task from outside stakeholders like providers and vendors.

Since you're managing a big extensive variety of clients and companies, you furthermore mght need to ensure you're handling those contracts the in the proper manner.

Stakeholder Management specializes in Taking care of stakeholders goals

It includes searching after the first-rate stakeholders worried at the challenge and looking after their goals and pastimes regarding the project.

Any unmarried mission can also have a massive wide variety of numerous stakeholders and each one in each of them need to have very extraordinary expectancies from the venture.

So, stakeholder control is all approximately the way you clearly meet the ones expectancies and the way are you able to have interaction with the ones stakeholders

The difference amongst knowledge areas and method businesses

So, you is probably questioning what's the distinction amongst manner companies and expertise areas?

To maintain it easy and easy, i would love you to think of way companies due to the fact the appropriate degrees that a assignment goes via all of the way from beginning to ultimate the mission.

On the other hand, take into account knowledge regions due to the reality the system which you may use so that it will manipulate the project.

Knowledge Areas are approximately specialized regions of knowledge which is probably used on challenge management subjects, whilst Process Groups examine that information.

While expertise areas offer facts on a particular area to help control a mission, the approach agencies are the manner via which that facts is finished on a venture.

Chapter 22: How to Manage Teams – Best Practices and Must Do's

In this phase I'm going to percentage my top suggestions for handling agencies. Once the mission starts offevolved offevolved, ideally speakme you need your institution individuals to usually perform like a well-oiled device with a easy reason with out getting derailed. These suggestions will assist you make certain your group contributors are capable of supply high productiveness, keep away from issues because of miscommunications/lack of expertise, guide every other and treatment issues quicker.

Develop Clarity of scope

Providing clarity in phrases of precisely what art work is meant to be finished on the undertaking can pass a protracted manner in growing crew effectiveness. Each character on the team desires to be truly smooth on:

•	What is the venture

•	What are the undertaking's desires?

- What desires to be accomplished to finish the assignment

Develop Clarity of Responsibility

Now that you've made it clear what the challenge is ready, it's far important to allow human beings recognize what their position is at the challenge. People want to recognize:

- What obligations are they liable for

- How can they accomplish the responsibilities they are responsible for

- Who are they running with and whom can they ask for help or guide

Develop Team member Onboarding and Problem Resolution Guidelines

Do you keep in mind the number one time you began out operating at your new undertaking? Everyone comes throughout demanding situations and troubles during art work, and counting on unique cultures, humans also can discover it embarrassing to ask for help. This is especially real for large organizations wherein new hires can specifically experience out of place due to the reality they've simply joined in and don't apprehend their manner throughout the vicinity or the humans.

Oftentimes, we take as a proper that people will artwork their manner round troubles. However,those sports activities can waste time and create undue strain. It might be an extended way extra beneficial to help your group member get their real challenge paintings achieved in place of set them on an adventure to find out in what branch does John from the crook group sit in order that they will get "X" file from him.

Before the beginning of every challenge, supply an extensive assessment for the economic organization enterprise, the maximum applicable sources they might need to engage with, the people will probable interact with, the files/materials they'll be probable to utilize, and function a detailed onboarding session together in conjunction with your business enterprise participants.

Get Buy in

Once humans are easy at the scope and obligations on the assignment, it's far important to get their Buy in. Getting humans to kingdom their Buy in for the task is corresponding to having them famend that positive we understand the venture, it's requirements, and we observe our roles and duties.

Develop Milestones

To music task development, the group wants to understand how a protracted way it has emerge as capable of study its universal overall performance.

Setting up every day/weekly/monthly milestones can help make your development obvious and help making a decision the manner to first-rate manage the challenge transferring ahead.

You can ask the subsequent to decide how properly you're doing at the undertaking collectively together with your group:

• What paintings has been completed and what remains left over, have we completed a enough percentage of the paintings as planned or are we lagging at the back of in shipping?

• Are we over finances or underneath fee variety/How an lousy lot have we spent, are we able to despite the truth that complete the challenge with the cash left

• Are we in advance of agenda or now not on time, How heaps time are we able to have left, are we able to complete it inside the final time

Establish Ground Rules

Ground policies are the easy mode of operation or pointers based totally totally on which the institution have to carry out. It is a code of behavior which the enterprise is supposed to abide through the use of. Ground regulations can help organizations keep employee pleasure and regular overall performance degrees at the same time as fending off unwanted behaviours/movements that can harm the task.

How conferences may be carried out

Setting ground hints for conferences can assist keep away from dropping time and making uninformed selections.

Ground hints can variety from manager to manager and from one enterprise culture to the opposite. I really have located the subsequent technique and flow to be useful:

• Specify the subjects of discussion and assembly time desk

• Each group member shares their points of debate beforehand to all assembly individuals

• All participants turn out to be aware of discussion factors (improvement updates from each member) and traumatic conditions being confronted

•   Actual assembly does no longer live on fundamentals and focuses as an alternative on answers to troubles

•   Meeting participants percent solutions and constructive comments

•   Repeat for subsequent meeting

Having an time table of debate elements distributed to all of the assembly members can ensure that everyone is aware about the fundamentals. The speak can then bypass ahead closer to greater efficient choice-making conversations.

Prohibiting the usage of cellular phones during indicates can make certain the contributors are respected and their factors are well understood.

Delivering positive comments in location of criticism is vital. Members of the assembly have to be endorsed to percentage thoughts on fixing problems in location of fostering hostility amongst each extraordinary. There absolutely is not any trouble in bashing a group member, it really contributes to creating the place of business extra poisonous. Instead, to location it in fact, it's higher to attention on in which we stand in recent times, what are we able to need to do to move wherein we want to head. With this technique, you may in reality reputation at the demanding situations a nice organization member is coping with and the participants can then endorse/offer aid on overcoming them.

Changes asked via the purchaser

Additional capabilities situation:

Customers can regularly request adjustments on projects. What is important to discern out is, whether or no longer the changes in shape into the prevailing scope or now not. Clients can also frequently request more competencies in case you need to now not be part of the genuine scope that changed into determined. Adding ON to the winning scope approach which you now have to skip decrease lower again and redo your baselines. Since the improvement of new abilties may also require additional art work, it is likely your scope baseline would encompass new deliverables, the time table will amplify because of the truth you currently require greater time, and your prices will growth due to the extra materials / skilled professionals you require to increase the ultra-modern-day function.

Remember, it's far crucial to have the patron agree in your preliminary deliberate scope, the purchaser need to renowned the deliverables which have been at the start planned. In case the client does demand more talents, they'll recognize their request goes over and above the real scope.

However, in case the customer's request is within the deliberate scope, then the ones functions are to be included without adjustments to the baselines. Walkthrough of the manner to make adjustments

Now allow's move over little by little a manner to position into impact changes on projects:

1. You gets maintain of a request to make adjustments on the assignment

Once you've acquired a change request, you currently want to get distinct information approximately the character and specifications of the alternate. It is important to apprehend what exactly is the person of the exchange being requested after which speak with the stakeholder to make certain you're on the equal web page. It is on your satisfactory hobby to get all the records in an expert Change request shape. Once you've amassed that statistics, it is always an first rate workout to yet again run that records thru the person who initiated the exchange to amplify a not unusual facts of what is required.

Most importantly, it's miles critical to ensure you get the initiators professional purchase-in and acknowledgement that sure this is the correct description of the kind of change they will be soliciting for. This will protect you from some of ability issues that might stand up due to miscommunication or misunderstandings.

2. Analyze the effect of the alternate at the mission

Next, it's miles essential to assess and look at the impact of the trade. You need to realise what will be the stop result of the effect on the undertaking constraints (the undertaking constraints embody Cost, Schedule, Scope, Customer Satisfaction, and Risk).

Most importantly, you've got to investigate the effect of the alternate on the triple constraint. The triple constraint consists of scope, fee and time table and those are taken into consideration the most critical constraints of the assignment.

3. Approval or Rejection of Change Request

After evaluating the change request for its effect, you could make the selection to both put in force or reject the trade. Changes are reviewed through the Change Control Board, do not forget it as a committee who is accountable for evaluating the trade and figuring out whether or not to put in force it or not. This committee can embody considered one of a kind critical stakeholders from the mission together with:

- Project Manager

- Experts

- Project Sponsor

- Functional Managers

- Customers

For small changes like adjustments, you have got the discretion as the venture manager to take a desire because the ones are nicely interior your authority. However, for big and higher effect modifications that might substantially trade the scope, fee range and time table, you could need to consult the Change manage board, especially the sponsor and patron.

Changes which can be inside the deliberate scope are greater with out trouble implemented, as an instance this could encompass changes made to the task to deliver it decrease once more on route.

In case of changes which might be out of scope e.G. Adding extra skills to a software program, the ones might probably require you to make top notch adjustments to the venture files.

Which leads us to our subsequent component-

4. Update the undertaking files

Once you acquire the exchange request, it's miles now time to record the ones changes. The mission files can be updated to mirror the trendy specifications of the challenge. The mission manipulate plan and baselines (scope, time table, charge) need to mirror the modifications, together with, what new artwork wishes to be finished, what's going to be the extra costs, and what sort of greater time is wanted to finish the project due to the exchange request.

In case the documents are not updated, human beings would possibly emerge as walking on distinct variations of the project that would result in in addition problems. The intention is to make certain that everybody is running at the same model of the task. Therefore, from this component onward, everybody can be walking on the revised model of the Project Management Plan and Project baselines.

five. Communicate and spread the phrase about the modifications

Now which you've up to date the challenge documents, it's vital to tell every body involved at the assignment regarding the modifications. Your task group, senior manipulate, and a few one of a kind relevant stakeholders need to be made privy to any updates to the challenge. If people are not noted of this communication, it may be adverse to the fitness of the assignment. So you want to help make certain they understand approximately what's taking place with a view to regulate make any required changes to their paintings as appropriate.

Chapter 23: What is a Project Charter – Define Project Charter Process

The Project Charter is a report that describes the whole venture. It gives you an entire high-diploma thoughts-set on what the venture is ready, and the whole thing associated with it. This is your Go To document with regards to coping with the task.

Contents of the Project Charter

The task charter consists of statistics regarding:

• What are Objectives and Goals

• What is the Business Case

• What is the Project Description

• What are the Project Requirements

• What is the Product Scope

• Who is the Project Manager

• What is the Success and Acceptance Criteria

- What are the Constraints

- What are the Risks

- What are the Milestones

- Who are the Stakeholders

- What are the Assumptions

- What is the Initial Budget

- What is the Initial Schedule

Develop Project Charter Process

A Project charter is advanced with the aid of way of figuring out stakeholders and discussing the number one statistics factors of the undertaking. The way to create a assignment charter is to talk to all the stakeholders who might be involved with the mission and from those discussions you will create the essential information factors of the mission. Then, because of the discussions with stakeholders, the contents of the challenge charter are superior.

Important matters to note approximately the Project Charter

• The first element to realize is that a assignment charter is a report which describes the task on an preliminary and typical diploma. What do I imply after I say the undertaking constitution describes the undertaking on initial and standard diploma? I'd collectively with you to undergo in mind the mission charter as a quick however cohesive and logical precis of all of the essential components of the project. Also be conscious that the project charter is written earlier than you in truth begin the undertaking.

• According to the PMBOK, The assignment charter authorizes the lifestyles of the mission. It also authorizes the project manager to use resources (monetary, human, cloth) to get the challenge finished. Simply talking, the challenge constitution makes the mission high-quality and brings it into life in order that paintings can begin at the assignment.

• The charter plans the challenge at a excessive degree, and consists of the essential elements of statistics of a task, it additionally consists of the Broader statistics of the mission

• The creation of a assignment charter is a manner to get the assist of the senior manage by way of approach of making them gather and decide to the existence of the challenge.

• A mission constitution includes the goals and dreams, what's the fulfillment standards, what are the immoderate-degree necessities, and what are the immoderate-degree risks and constraints the task is predicted to face, what assumptions are we making and so forth.

• Anyone who reads the undertaking charter have to right away be capable of get an idea what the task is ready and what are the most critical subjects to recognize about it.

• It's vital to realize that the Project Charter is not a report that includes in depth detail, rather, the project is planned in detail best after the signing of the undertaking constitution. The reason that the constitution does now not contain loads of element is due to the reality unique making plans takes extra time and costs more money, so targeted planning great starts offevolved as soon as the assignment constitution gets signed thru senior control.

• Furthermore, we need to hold in mind that The Project Charter is signed by way of the sponsor of the undertaking or by using the usage of senior manage.

• Also, have a examine that the challenge supervisor may also or won't create the challenge constitution. The sponsor can also ask the assignment supervisor to create the constitution or the constitution can also already have been created in advance than the task manager is introduced on board to manipulate the project.

Project Charter Template

Here is a pattern template for a Project Charter. I'd like you to don't forget the task you're walking on and attempt filling out the information in the remarkable sections.

Business Need/Business Case – (Why is the Project required)

Project Description (What is the Project about, How will or now not it is achieved)

Project Scope (What artwork is needed to do the mission)

Key Deliverables (What can be the outputs of the challenge)

Project Objectives (What key matters are we able to assume to benefit, What is the Success Criteria)

Stakeholders Who may be concerned inside the venture and who is probably affected by it

Budget (How masses can we count on the Project to Cost)

Schedule (How plenty time would it take to finish the assignment)

Milestones (What are the key measurable achievements and improvement dreams of the undertaking)

Constraints (What can restrict development on the venture)?

Risks (What High degree Risks can the Project face)

How to Perform Integration Management

Now which you've discovered about a way to broaden the project charter, let's flow into immediately to the next machine of integration control, it simply is the way to extend the project manage plan.

Develop Project Management Plan Process

The undertaking manipulate plan is a report which incorporates all of the character control plans along aspect the baselines of the product. So, what precisely does that suggest? When it involves the unique information regions, permit's maintain in mind them to refresh our reminiscence:

- Integration Management Plan

- Scope Management Plan

- Schedule Management Plan

- Cost Management Plan

- Quality Management Plan

- Resource Management Plan

- Communications Management Plan

- Risk Management Plan

- Procurement Management Plan

- Stakeholder Management Plan

What is a Management Plan?

All of these taken into consideration one of a type control plans include the technique for managing the techniques of the task. So as a challenge supervisor you need to expand a separate control plan for every of these information areas.

Now what precisely are the best-of-a-kind sections of a control plan. There are three essential sections you need to bear in mind.

Sections of a Management Plan

- Planning

- Executing

- Controlling

Planning

First, allow's talk about Planning. The Planning segment of a control plan defines the techniques and techniques which can be to be followed for a selected understanding region. For instance, the making plans segment of the awesome manage plan will assist you to recognise what are the simplest-of-a-kind techniques and strategies that you want to have at the project in order to make certain excellent at the assignment.

Similarly, on the subject of the making plans phase of our danger control plan, it'll can help you understand about the amazing strategies and methods which you want to make certain on the challenge to control risks on the challenge.

Executing

Next is executing. The executing section will let you know approximately how precisely you are speculated to bypass about doing the methods and techniques which you described at some point of making plans. So to keep it simple, remember it this way, The planning segment is going to inform you about the "What" and the executing phase goes to tell you about the "How".

When it includes making plans excellent, it's going to permit you to recognise what you're speculated to do on the project so one can make certain satisfactory, and the executing segment is going to tell you strategies precisely you may enforce the plan and the techniques and the strategies to position into effect great.

Controlling

Lastly, allow's communicate the controlling section of a control plan. Controlling will inform you a way to test the overall overall performance of the mission. It may can help you understand approximately a way to select out out variances, e.G. In case you be conscious the venture goes off course, how will you deliver the overall performance again at the right tune.

That wraps up our dialogue approximately Management Plans. To summarize, genuinely remember that the Project Management Plan is a set of all of the person control plans for the particular data regions.

Direct and Manage Project Work Process

This is the way in which the assignment manager goes to get paintings completed on the task. These are duties that you may be involved in doing at some point of the venture. E.G. This consists of generating deliverables, this means that you're getting the paintings achieved through the crew and they may be going to be developing the deliverables of the project.

You also can be involved in amassing information from all of the mission art work that is taking location round you to ensure you preserve song on its development. Furthermore, you may be involved in getting adjustments achieved on the task in case they'll be required or requested. Lastly, you'll be dealing with all the particular organization participants and the stakeholders who're involved on the undertaking.

Manage Project Knowledge Process

As a undertaking manager you're answerable for continuously making upgrades to the task so one can make certain its achievement. Furthermore, you moreover may additionally need to adopt steps so you can avoid failure.

Management of the know-how that is created at the project is a totally crucial thing in phrases of the duty of the challenge manager.

Firstly, you can use the knowledge from beyond initiatives so that you can manual the execution of your cutting-edge and destiny responsibilities. Also, have discussions with the stakeholders or your project group humans on how they worked at the preceding projects which have been of a comparable nature to the simplest that you're walking on right now.

You ought to use the knowledge and learnings from the preceding initiatives. Talk to your institution people on how they executed the tasks in the beyond. Researching information from past initiatives and from specific stakeholders can assist guide you in developing the execution plan for your contemporary-day and future obligations.

Furthermore, it's also your obligation to create know-how regarding your destiny and your cutting-edge-day-day tasks. Consider all of the matters which are taking region, the conflicts and troubles that you're dealing with, the opportunities which you're capitalizing on, and the good stuff that you're executing on the mission. All those elements have to be considered and recorded in order that a person within the destiny can employ them. Therefore, you may gather the expertise, practices, and records concerning your modern-day venture and document it, and that's what Managing Project Knowledge is ready.

Monitor and Control Project Work Process

As Project Manager, you ought to be tracking, reviewing and reporting the development of the assignment on a non-forestall foundation. This is because of the fact it's miles very vital to evaluate the modern-day and the forecasted performance of the task to the overall ordinary overall performance which became deliberate inside the direction of the planning segment. This evaluation can tell you whether the venture's performance is heading in the right path or now not. Hence, it will can help you answer whether or no longer your planned scope, the fees you expected, and a while desk is on track.

Perform Integrated Change Control Process

(*Please be conscious that there may be additionally a separate trouble depend on a way to make modifications on a assignment which covers the entire system in hundreds more element. In this segment, I will summarize it greater in brief.)

Next is the carry out included trade manipulate manner. However, first we need to find out approximately what exactly is a exchange request. When you are going for walks on any challenge whether or no longer it's miles a production associated undertaking or a era associated mission there are sure to be severa changes as a way to expose up for the duration of the route of the project. As the venture manager, it's far your duty to govern the changes that show up on a project.

Change requests discuss with a right request for changes or modifications which want to be made at the project. Changes may be started out thru your client, by using manner of your crew individuals, or via way of the use of your self if you note the venture calls for them. The Perform blanketed alternate control device is the formal technique in which change requests are evaluated and then they may be each commonplace or rejected. It isn't crucial if you want to take shipping of each exchange request that comes via. For any change request which you purchased you want to evaluate them for the effect that they may have on the venture. The impact of a exchange request needs to be evaluated on regions which includes:

- Cost

- Schedule

- Quality

- Risk

- Resources

- Customer Satisfaction

For instance, you need to preserve in thoughts what will the impact of the price on my venture if the exchange request is executed. Are my fees going to increase? Is the assignment going to take extra time to finish? Is it going to lower or increase the great? Are there any dangers of implementing that exchange request? Hence, you want to recall a couple of components earlier than you without a doubt implement a trade request and popular.

Close Project Process

This is the method wherein you may entire and wrap up that mission. Now permit's communicate about what does the venture manager must do to close the venture. Firstly, it's far very essential that you make sure the paintings has been finished in step with the necessities of the assignment.

Secondly, you want to get the formal popularity of the challenge or the product from the client. The formal recognition manner you show the very last deliverable to the client and note whether or not or now not they take transport of it or not. At this factor, if the customer accepts, this suggests the deliverable has been formally finalized and stated via the patron.

Also,It is vital to complete the financial reporting on the project and to complete the closure of the procurements on the purchases that you have made. You will gather and archive the mission statistics and the know-how made from the assignment in order that they'll be used for reference in future obligations. You can even gather and update the final commands which you studies. You may also deliver the completed product to the customer. This isn't similar to getting the formal recognition from the customer because within the first case you are becoming the acknowledgement from the client that they acquire this product. After that, you moreover may additionally moreover want to make sure which you're turning in the finished product to the purchaser for the final handoff. And sooner or later, you will get comments from the consumer regarding the project.

Note that every task is closed, no matter whether or not or now not or no longer it is completed, or possibly if it is left incomplete and halted due to some motive, you need to wrap up and close to the challenge down officially.

Chapter 24: Scope Management

What is Scope Management

The first critical factor for us to apprehend is, what does scope propose in terms of project manipulate?

Simply speaking, Scope refers back to the artwork this is required to be achieved on the assignment.

Now allow's talk about what is the reason of the Scope Management.

Scope Management is all approximately ensuring that the art work executed at the project includes simplest the paintings required to complete the challenge and then ensuring pleasant that art work is accomplished and not anything greater is carried out on the task.

Therefore, we are able to say that Scope Management basically entails matters:

•   Firstly, figuring out what shape of art work is needed to be completed at the project.

•   and Secondly, ensuring best the paintings required is accomplished on the project.

Understanding Scope Further

First, we will apprehend how scope is evolved on the assignment

At the start of the challenge all of the necessities are amassed from all the stakeholders who're involved on the venture. These necessities are then prioritized as per the industrial organization case and the challenge necessities.

The Prioritization of requirements consequences in determining what artwork is covered in the project and what art work isn't covered inside the venture i.E. To determine what's inner scope and what's out of scope.

The scope is then permitted officially in advance than the paintings begins offevolved. By normal formally we suggest to say you're making it clean to all people on foot at the venture as to what art work is needed to be finished to finish the undertaking. After that, you get a log off to get their acknowledgement/buy in.

The scope of labor is then represented through a WBS (Work Breakdown Structure). You can don't forget it as an intensive map of the task work or an define of all of the art work to be completed at the mission. This substantially permits make clean the scope worried of the assignment.

Important notes you need to recognize approximately dealing with scope:

Changes to scope are not effects allowed and can handiest be made with an approved change request. (This is because you positioned a scope baseline within the path of planning, and the undertaking is predicted to have a look at this baseline in some unspecified time in the future of its length. Therefore, changing any baseline want to be considered a completely remaining hotel preference.)

Secondly, we ought to go through in thoughts that Changes to scope are first analysed for their impact on mission constraints, which include costs, time, danger, property, patron delight and first rate.

Another component to phrase is that Determining what is included and what isn't always blanketed inside the scope ought to be a non-save you way with the resource of the usage of the Project Manager. This facilitates high best the undertaking remains on route.

And in the end, Changes to scope that do not in form with the Project Charter must not be prevalent.

Scope Management involves the following tactics:

• Plan Scope Management

- Collect Requirements

- Define Scope

- Create WBS

- Validate Scope

- Control Scope

What are Project Requirements

Project requirements communicate to the outline and functions of the services or products being developed. The real necessities of a undertaking embody the unique scope, i.E., the info of the artwork to be completed on the project. Requirements are amassed at a High Level in the Project Charter at the begin of a assignment, these are then fleshed out to enlarge the precise scope.

What are Constraints

A Constraint is any detail that limits your functionality or your options of doing the assignment.

An smooth clarification would be to reflect onconsideration on constraints due to the reality the variables or the topics that could prevent you from doing the mission.

Constraints are recognized finally of venture beginning at a excessive stage. Then, During Project Planning, constraints are detailed inside the Define Scope Process. This way that constraints are recognized from the pinnacle during the start level, however it isn't until the outline scope technique that constraints are elaborated in element and further statistics is collected and understood approximately them.